Cyber Technological Paradigms and Threat Landscape in India

Ramnath Reghunadhan

Cyber Technological Paradigms and Threat Landscape in India

Ramnath Reghunadhan
Department of Humanities and Social
Sciences
Indian Institute of Technology
Madras
Chennai, Tamil Nadu, India

ISBN 978-981-16-9127-0   ISBN 978-981-16-9128-7   (eBook)
https://doi.org/10.1007/978-981-16-9128-7

This Palgrave Macmillan imprint is published by the registered company Springer Nature
Singapore Pte Ltd.
The registered company address is: 152 Beach Road, #21-01/04 Gateway East, Singapore
189721, Singapore

Defeat is a state of mind. No one is ever defeated until defeat has been accepted as a reality
—Bruce Lee

The book is dedicated to my family, teachers and friends who loved, cared and guided me through all hardship, and gave me strength and direction in life. This book is not just an assemblage or compilation of knowledge and analysis, but a remembrance of "where, what and how" life changes you.

Acknowledgements

I am thankful to several individuals who have encouraged and supported me in the completion of this work. Firstly, my deepest gratitude to my family, Mr. V. Reghunathan, Ms. Suma Reghunathan, Ms. Lekshmi S. Nath, Mr. Krishnanath Reghunadhan, Mr. Chandan Shibu and other family members for their wishes and support. The inspiration and support provided by my teacher and supervisor, Dr. Joe Thomas Karackattu, Associate Professor, Department of Humanities and Social Sciences, Indian Institute of Technology Madras, India has been monumental during all these times.

I would like to express my special gratitude to Air Marshal Vinod Patney, Centre for Air Power Studies (New Delhi, India), Dr. E. Dilipraj, Research Fellow, Centre for Air Power Studies (New Delhi, India); Mr. Starmon Pillai, Inspector, Hi-Tech Cell, Police Head Quarters (Kerala, India) for their insights on the topic. My gratitude to the academic, scholarly and scientific community of CyFy 2017 organized by the Observer Research Foundation (ORF), particularly Dr. John Mallery, Research Scientist, Computer Science and Artificial Intelligence Laboratory, Massachusetts Institute of Technology (MIT) (USA); Dr. Rahul Misra, Senior Lecturer, University of Malaya (Malaysia); Dr. Rakesh Mehrotra, Vice President, International Telecommunication Union- Asia-Pacific Telecommunity (ITU-APT) Foundation of India (New Delhi); Mr. T. George-Maria Tyendezwa, Head of Cybercrime Prosecution Unit, Federal Ministry of Justice (Nigeria); Dr Aadya Shukla, Research Scientist,

Computer Science & Artificial Intelligence (AI) Lab, Massachusetts Institute of Technology (MIT) (USA); Dr. Dennis Broeders, Research Fellow, Netherlands Scientific Council for Government Policy (Netherlands) and Dr. Chuanying Lu, Senior Fellow, Institute for Global Governance Studies (China) for providing academic inputs, support and encouragement for the completion of the work. More importantly, I would like to thank the Editor and reviewers of Springer Nature and Palgrave Macmillan for their extensive review and comments on the book.

I am also grateful to the faculty, librarians and staff of the School of International Relations and Politics (SIRP) and Mahatma Gandhi University Library. Further my sincerest gratitude to my teacher Dr. Lirar Pulikkalakath, Assistant Professor, School of International Relations and Politics, Mahatma Gandhi University, for his guidance, inputs and support for the work. I would also like to express my sincere thanks to Prof. A. M. Thomas, Prof. K. M. Seethi, Dr. C. Vinodan, Dr Suresh Kumar, Dr. Vimal Kumar Vazhappally, and Savithry T. K. all other teaching and non-teaching staff of the University for their timely advice and encouragement. Moreover, I would like to thank Dr. C. Indira Devi (T.K. Madhava Memorial College), Prof. K. Indulekha (Mahatma Gandhi University), Dr. Ashish Mahabal (California Institute of Technology), Prof. J. G. Ray (Mahatma Gandhi University), Dr. Ren Meghanad Gupta (Cotton University), Dr. Anish K. Santhipuram (K.S.M. Devasom Board College), Prof. Suresh Rangarajan (Kerala University), Dr. Chandra Rekha (Indian Council for World Affairs), Prof. Sajad Ibrahim K. M. (Kerala University), Prof. Joseph Antony (Kerala University), Prof. Shaji Varkey (Kerala University), Prof. Gireesh R. Kumar (Kerala University), Dr. Samuel J. Kuruvilla (Kerala University) for their support all along. A special thanks to the support, cooperation and help extended to me by friends Ms. Aiswarya Unni (Mahatma Gandhi University), Mr. Ansel Elias Stanley (Jawaharlal Nehru University), Mr. Rahul Raj (Kerala Media Academy) as well as other friends and well-wishers.

Ramnath Reghunadhan

CONTENTS

About the Author

Ramnath Reghunadhan is a Research Scholar at the Department of Humanities and Social Sciences, Indian Institute of Technology Madras (India), and is an alumnus of School of International Relations and Politics (SIRP), Mahatma Gandhi University (India). Recently, he was also the recipient of the *Mira Sinha-Bhattacharjea Award* (MSB 2019) by the Institute of Chinese Studies (India), and the 2021–2022 *Huayu Enrichment Scholarship* by the Ministry of Education of Taiwan (Republic of China). The author is a reviewer at *Sage Open* (USA), *International Migration* (UK), Common Grounds Research Networks (USA), ASTES Publishers (USA). His professional experiences include Research Internship at the Center for Air Power Studies (CAPS) (India), and is currently a member of Institut de diplomatie publique (UK), and Globelics (Costa Rica). His papers have been indexed in Stanford Libraries (published by Stanford University), Periodical Articles for Current Awareness (published by US Army War College Library) and Current Awareness Bulletin (published by Centre for Women Development Studies). Recent publications have been in journals like *Asian Affairs* (Royal Society of Asian Affairs), *Journal of Asian Security and International Affairs* (Sage), *Journal of Black Studies* (Sage), and *Strategic Analysis* (IDSA). He has published book chapters in *IoT and Analytics for Agriculture* (Springer), and *Handbook of Research on Blockchain Technology* (Academic Press). The author was also part of the *Track I public consultation* for India's Science Technology and Innovation Polity (Ideathon Vision 2020) by

xvi   ABOUT THE AUTHOR

the Department of Science & Technology (Government of India) and Science Policy Forum, and the contribution was selected among *Top 50 Ideas* for India's Science Technology and Innovation (STI) Policy. Research interests include Technology in International Relations, STI-related policy and governance, political economy, China Studies, political economy, international security, research methods, empirical research, positivism, technology in peace and conflict studies, global governance and regionalism.

ABBREVIATIONS

AI	Artificial Intelligence
AIDS	Acquired Immuno Deficiency Syndrome
ALICE	Artificial Linguistic Internet Computer Entity
ATM	Automated Teller Machine
AWS	Autonomous Weapons Systems
BBN	Bolt, Beranek and Newman
BBNL	Bharat Broadband Network Limited
BCI	Brain-Computer Interface
BE-ID	Blockchain Emergency Identity
BSE	Bombay Stock Exchange
BSNL	Bharat Sanchar Nigam Limited
CAN	Controller Area Networks
CCA	Controller of Certifying Authorities
CCTNS	Crime and Criminal Tracking and Network Systems
CCTV	Closed-Circuit Television
CD	Compact Disk
CDAC	Centre for Development of Advanced Computing
CEO	Chief Executive Officer
CERL	Computer-Based Education Research Laboratory
CERT-In	Indian Computer Emergency Response Team
CIH	Cheng Ing Hau
CIIP	Critical Information Infrastructure Protection
CNI	Critical National Infrastructure
CNN	Cable News Network
CSCs	Common Service Centres
CSIS	Centre for Strategic and International Studies

CSK	Cyber Swachh Kendra
CyAT	Cyber Appellate Tribunal
DARPA	Defence Advanced Research Projects Agency
DDoS	Distributed Denial of Services
DeGS	District e-Governance Society
DILRMP	Digital India Land Records Modernization Programme
DNS	Domain Name System
DoP	Department of Posts
DoS	Denial of Service
DoT	Department of Telecommunications
ENISA	The European Union Agency for Network & Information Security
ERNET	Education and Research Network
ESDM	Electronics System Design and Manufacturing
EUROPOL	European Union Agency for Law Enforcement Cooperation
FBI	Federal Bureau of Investigation
Gbps	Gigabyte Per Second
GDP	Gross Domestic Product
GoI	Government of India
GPR	Government Process Re-engineering
GUN	Government User Networks
H2M	Human-to-Machine
I4C	Indian Cyber Crime Coordination Centre
ICT	Information and Communication Technology
IDPs	Internally Displaced People
INTERPOL	International Criminal Police Organization
IoP	Internet of People
IoT	Internet of Things
IPC	Indian Penal Code
IPT	Intelligent Public Transport
IRC	Internet Relay Chat
IRINN	Indian Registry for Internet Names and Number
ISTR	Internet Security Threat Report
IT	Information Technology
LDCs	Least Developed Countries
MCA	Ministry of Corporate Affairs
MeitY	Ministry of Electronics and Information Technology
MHA	Ministry of Home Affairs
MIT	Massachusetts Institute of Technology
MLAT	Mutual Legal Assistance Treaties
MMPs	Mission Mode Projects
MNCs	Multi-National Companies
MoC	Ministry of Communication

MoD	Ministry of Defence
MSME	Micro, Small & Medium Enterprises
MTNL	Mahanagar Telephone Nigam Limited
NATGRID	National Intelligence Grid
NATO	North Atlantic Treaty Organization
NCCC	National Cyber Coordination Centre
NCIIPC	National Critical Information Infrastructure Protection Centre
NCRB	National Crime Records Bureau
NeGP	National e-Governance Plan
NGIS	National Geo-Spatial Information System
NGOs	Non-Governmental Organizations
NIB	National Information Board
NIC	National Informatics Centre
NICSI	National Informatics Centre Services Inc.
NIELIT	National Institute for Electronics and Information Technology
NII	National Information Infrastructure
NIXI	National Internet Exchange of India
NKN	National Knowledge Network
NMEICT	National Mission on Education through Information and Communication Technology
NOFN	National Optical Fibre Network
NPE	National Policy on Electronics
NRIM	National Rural Internet Mission
NSA	National Security Agency
NSDG	National e-Governance Services Delivery Gateway
NSE	National Stock Exchange
NSSO	National Sample Survey Organisation
NTRO	National Technical Research Organisation
NWs	Networks
OS	Operating System
P&G	Procter and Gamble
PC-DOS	Personal Computer-Disk Operating System
PDS	Public Distribution System
PLATO	Programmed Logic for Automatic Teaching Operations
PMO	Prime Minister's Office
R&D	Research and Development
SCA	Swiss Cracking Association Virus
SCADA	Supervisory Control and Data Acquisition
SDCs	State Data Centres
SLI	Special and Local Laws
SMS	Short Message Service
SNAP_R	Social Network Automated Phishing with Reconnaissance
SQL	Structured Query Language

SSDGs	State Service Delivery Gateways
STQC	Standardisation Testing and Quality Certification
SWAN	State Wide Area Network
Tbps	Terabytes Per Second
TCP/IP	Transmission Control Protocol or Internet Protocol
TDSS	Tommy Douglas Secondary School
TERM	Telecom Enforcement Resource and Monitoring
TFN	Tribe Flood Network
TRAI	Telecom Regulatory Authority of India
UIDAI	Unique Identification Authority of India
UIUC	University of Illinois at Urbana Champaign
UK	United Kingdom
US DoD	United States Department of Defence
USA	United States of America
USB	Universal Serial Bus
USD	United States Dollar
VSAT	Very Small Aperture Terminal
Wcry	Wannacry
WFP	World Food Programme

LIST OF FIGURES

CHAPTER 1

Introduction

Abstract The chapter contextualizes and conceptualizes various aspects in the field of cyberspace and the related threat landscape.

Keywords Cyberspace · Cyber threats · Dimensions · Cyberwarfare · Cybercrime · Cyberespionage · Cyberterrorism · Digital India

INTRODUCTION

The term cyberspace originated and evolved as a conceptual understanding from the science fiction works of William Gibson, namely *Neuromancer*. The work defined cyberspace as "the matrix" (Gibson, 2004). Gibson's works were three books in total; the other two were *Count Zero* (1986) and *Mona Lisa* (1988). In *Neuromancer*, the protagonist hacker dreamt of being in cyberspace but ultimately has to face off against a powerful entity powered by artificial intelligence (Gibson, 2004, 2). The description of cyberspace in the book provides similarities to what is seen in contemporary times, wherein cyberspace is considered as a "consensual hallucination" that sustains through the interconnected nature of a large computer network. Accordingly, cyberspace is

R. Reghunadhan, *Cyber Technological Paradigms and Threat Landscape in India*, https://doi.org/10.1007/978-981-16-9128-7_1

A consensual hallucination experienced daily by billions of legitimate oper-
ators, in every nation, by children being taught mathematical concepts... A
graphic representation of data abstracted from the banks of every computer
in the human system. Unthinkable complexity. Lines of light ranged in the
nonspace of the mind, clusters and constellations of data. Like city lights,
[they were] receding... (Ibid., 42)

These concepts influenced a lot of classic science fiction movies, especially the classics *The Matrix Trilogy* (1999–2003). This movie depicted
cyberspace as the creation of an "interconnected system" in the future,
whereby nothing is devoid of being connected to "The Matrix". This
shows where human privacy is a mythology, and more importantly,
decision-making and free are just stories and lies created to suffice human
oppression and exploitation through (un)conscious consensus of the
oppressed (Bell, 2009, 468–471; Goodreads, 2018).

The world is considered as becoming flatter and smaller. A connotation
used in this stead is the "shrinking of the world" due to the accelerated
advances in technologies related to information and (tele)communication.
This has also led to issues from the physical world perpetuating into the
digital world as well. The "information umbrella" has also led to the rise
of hackers, which have shown increasing threats at both the individual
and state-levels as well. A computer or a network of computers are often
attacked and hacked by malicious software and tools. This helped gain
control over infected devices and systems. This ranged from impacting
an individual's details and assets to nuclear plants, industrial production centres, and countries' power grid to international organizations'
decision-making channels (Cobb, 2014).

Globally, this is an era of digitalization, (inter)connected through technological paradigms that have created new avenues and opportunities for
human civilization advancement. Further, this created and produced new
spaces and spheres of interaction and are often construed as cyberspace,
virtual or digital world. These have also led to transformation and transition in global politics and international relations, especially in military,
diplomacy, sovereignty and globalization. The contemporary times is
generally perceived as the period of digitalization, which includes influences and paradigmatic shifts in the physical world. There are impacts on
social, economic, political and biological aspects of human existence and
well-being (Davis, 2016; Dufva & Dufva, 2019, 17–28).

Beginning with the latter half of the twentieth century and two decades into the twenty-first century, the world has seen the increasing, multiplier and exponential strides in the form of digital inter- & intra-connectivity, with advancements in the internet revolution, information and communication technologies (ICTs), telecommunications, artificial intelligence, machine learning, deep learning, manufacturing, production technologies, financial technologies, cyber governance and a plethora of other technologies across the world (Malik, 2004, 102–120; Schwab, 2016). During the aftermath of the 9/11 attacks, Jacques Derrida, the French philosopher, argued that the cyberthreat is "a more potent threat to our political and physical world, [changing the] psychological and historical sense of a violent attack, and the concept of territory" (Schweitzer et al., 2013, 23). Cyberthreats are considered to be occurring in the variant spatial paradigm, called cyberspace, but integrated into the physical world.

CYBERSPACE AND SOCIETY

Gibson's work was more in line with futuristic applications and analogues of cyberspace, with implications that have led to sociocultural and political implications later on (Featherly, 2021). The emergence of cyber technologies threats has increased the relevance of geopolitical, socio-economic, legal and ethical relations and activities (Lagarde, 2018; MeitY, 2019). With this enhanced relevance for the number of socio-political activities, international relations and the related power politics have been greatly influenced by cyberthreats. The redefinition of politics with variant facets of interrelationships arises, determining and reshaping the "sphere of influence" of many dominant countries; conventionally in the physical world (SIPRI, 2020). The increasing relation concerning physiological and psychological factors will transform the domain of national security and human security.

There is a transformation of what rights are, being redefined by the traditional-cum-conventional actors and/or institutions, within both the physical and the virtual world. The conceptualization of security has to be redefined based on the emerging needs and threats that have cyberspace part of being inherent in it. The changing global political economy in itself often problematizes the conventionally perceived upon frameworks of international security, economy and socio-political organization, including the spread and development of information, data, ideas and/or knowledge (Brown, 2014; Maher, 2013). In 1996, John Perry Barlow famously brought out "*A Declaration of the Independence of Cyberspace*", where he asserted that cyberspace is a realm that is well beyond conventionally perceived and demarcated borders of the physical world (Barlow, 1996). He famously said that

> Governments of the Industrial World, you weary giants of flesh and steel, I come from Cyberspace, the new home of Mind... You are not welcome among us. You have no sovereignty where we gather... In China, Germany, France, Russia, Singapore, Italy and the United States, you are trying to ward off the virus of liberty by erecting guard posts at the frontiers of Cyberspace. These may keep out the contagion for a small time, but they will not work in a world that will soon be blanketed in bit-bearing media... We will create a civilization of the Mind in Cyberspace. May it be more humane and fair than the world your governments have made before. (Ibid.)

Transformations of Cyber Threats

According to the National Research Council report "The modern thief can steal more with a computer than with a gun" (NRC, 1991). It could be concentrated on a particular target continuously, consecutively and simultaneously. This was possible due to the mercurial developments in technology and innovation in the cyber domain. The systems and networks that operate in cyberspace have vulnerabilities emerging and significant risks to both individual organizations and national security. The ability and the utility of the governance, security and defence in cyberspace of any nation, including that of India, depend primarily on the capability, understanding and know-how of humans to deal with issues and situations in the cyber domain (Hall, 2018, 9–12).

The development of sensors, processors, monitors and its integration into everyday goods as well as the consumers, all of which are closely linked with both the virtual and physical worlds, have but transformed the relations. This also increased the quantity of data available and being produced. This has led to "the development of the Internet of Things (IoT)" and Big Data, which will further transform the nature of cyberspace, often transcending the importance of the physical world (Schwab, 2016). Haly Laasme once stated that:

> As anyone can infer the area of cyber defense and security; [which] is still profoundly [an] uncharted territory and its doctrine far from empirical

realisation. [There aren't] any internationally accepted definitions... What one nation considers a cyber attack might appear more like a [cyberwar] to another or even a simple [cybercrime] to a third (Laasme, 2011).

This transformation often alters the relationship between governments and citizens; it enables relatively low-cost communication platforms at light speed, thereby changing its governance and security aspects. These things can also create accountability and transparency at a greater level, particularly with the emergence of open data sourcing by different nations, institutional actors, groupings, civil society as well as other non-governmental organizations (NGOs). Unlike the conventional elitism or access to the top echelon of the society, a substantial number of the population have been part of this new virtual domain often, if not done routinely. This connectivity among computing and communication devices has led to the creation of cyberspace "as a single, shared virtual domain" (Reardon & Nazli, 2012). Based on the statement by the US Department of Homeland Security, cyber-related attacks create vulnerabilities in cyberspace, creating vulnerabilities "to steal information and money and are developing capabilities to disrupt, destroy, or threaten the delivery of essential services" (DHS, 2020a).

DIMENSIONS OF CYBER THREAT

The inherent nature and impact of the cyberthreats have so far transcended any normative idea of Westphalian boundaries that exist in the physical space–time paradigm. Cyberthreats concerning the territory of countries and/or modern states do not seem to apply to the conventional idea of sovereignty in the physical world. This was attributed towards the acceptance of physico-political boundaries. However, "the emergence of Transmission Control Protocol" or "Internet Protocol" (TCP/IP) transformed the very nature of the exchange of information or data. This laid the foundation of "digital globalization" (McKinsey & Company, 2016) or "digitalization of globalization", with each technological revolution in the field of ICT leading to transitions around the globe. Thus, the world is considered to have become smaller due to the advances in technology or the "shrinking of the world" due to the information umbrella (Kirsch, 1995). Overall, there are four dimensions of cyberthreats: cyberwarfare, cybercrime, cyberterrorism and cyberespionage (Fig. 1.1).

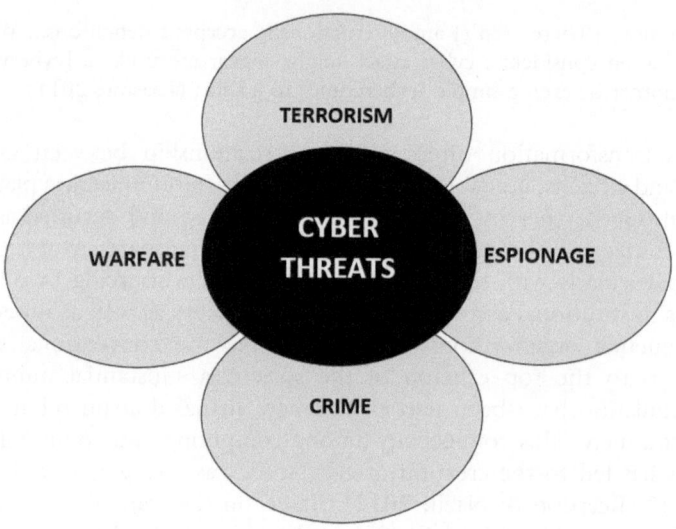

Fig. 1.1   Four dimensions of cyberthreats (*Source* Compiled by the Author)

The variant transformative threats in the form of cybercrimes, cyberwarfare, cyberterrorism, cyberespionage, intrusion and other forms of attacks from "State and non-State actors" are but testing the resilience of different nations in cyberspace. The number of threats to individuals, groups, organizations and/or states has seen a rise in the cyber domain, often leading to the need for the conceptualization of cybersecurity. The conventional threat perception of different actors in the cyber domain ensues a different scenario, unlike the strategic considerations of land, sea, air and outer space. The securitization issues have been but pertinent since the emergence of the cyber domain, often characterized as the "fifth domain" or the "fifth commonspace" (The Economist, 2010; Kumar, 2014).

The first dimension of threat in cyberspace, i.e., *cyberespionage* or cyber spying, is a type of malicious activity to covertly access and obtain information or knowledge in cyberspace conducted by the State actors as well as corporates. Throughout human civilizational history, beginning from tribes to nation-states, humans have mostly tried to undermine each other, especially through clandestinely, that focused on activities like spying, sabotage and/or subversion. Cyberespionage is more of a variant form of espionage activities that states historically undertook. It helps even extend and/or improve the power of a State in relation to other State actor(s). It includes cyber-related attacks done under the control and/or direction of the State, especially targeting other government(s) and/or private business(es) to steal information (particularly sensitive information) for political, commercial, military and/or strategic gain.

Cyberespionage has been conducted under the aegis of various state actors in enhancing its offensive strategies, capabilities. Further, it can negate rival countries and non-state actors (Rubenstein, 2014). Meanwhile, when a corporate undertakes espionage activities, it comes under the ambit of industrial or commercial espionage. The emergent interconnectivity through the new domain has but have transformed espionage activities undertaken and have potentially exacerbated the situation through greater strides in the platform of science, information and communication technologies. In 1982, the CIA (Central Intelligence Agency) of the US was supposedly said to have caused the gas pipeline explosion in Siberia by beguiling Soviets into stealing flawed industrial control software[1] causing the pipeline to malfunction (O'Brien, 2017).

In 1986, the first documented case of cyberespionage was the systematic targeting and hacking of computers at the US military bases. The hacker was "identified as Markus Hess, a West German national" who allegedly stole information for the Soviet intelligence organization, KGB (*Komitet Gosudarstv ennoy Bezopasnosti*) (Baines, 2011). In 1999, the Moonlight Maze case, where an FBI (Federal Bureau of Investigation) found out that different sectors of the US government was targeted and compromised and documents were stolen (O'Brien, 2017). The Duqu worm (in 2011), and the Flame (in 2012), were developed primarily for cyberespionage activities. Flame focused on users of infected computers, stole data as well as information in the form of "documents, recorded conversations and keystrokes". It also opened backdoor(s) in these "infected systems to allow the attackers to tweak and add new functionalities" (Zetter, 2012).

The reports of hackers stealing information sometimes allegedly supported by State actors like China, particularly against the US as well as against other "State and non-State actors", have become more of a norm in the current era. Various scholars believe that a substantial

---

[1] The ICS is a precursor to those used in SCADA systems.

quantity of resources has been deployed by China to spy on the government institutions and private companies in other nations (CNBC, 2012; Maizland, 2020), within and outside their respective boundaries. This transcending nature of espionage activities by the State actors is but blurring the distinctiveness of even commercial espionage and securitization measures, steps as well as activities. One of the game-changing events was the discovery of *Stuxnet* by Roel Schouwenberg of Kaspersky Lab in 2010. It gained much attention due to its use of zero-day vulnerabilities in spreading worm via removable drives, which was effective in overcoming air-gapping (Zetter, 2014).[2]

It was "designed to target industrial control software systems", with particular direction and capability to hijack a particular type of "Programmable Logic Controllers (PLCs) used to run and monitor industrial systems in the Iranian uranium enrichment program". It caused system malfunction and essentially sabotaged the systems in place (Kushner, 2013). Currently, cyberespionage is not just the activities of state and state-sponsored groups. In 2015, information of a group called *Butterfly* that compromised (a large) number of major corporates over the past three years was uncovered. It stole large-scale data, information and intellectual property of the corporates and sold it for the largest bidder or the contract provider (O'Brien, 2017). Countries and corporates lost terabytes of data from their systems (or individuals) due to espionage activities, particularly focusing on susceptibility and, most importantly, on the human-error induced predicaments.

The second dimension of threat in cyberspace; i.e., *cyberterrorism*, aims to target the fear of the population and creates fissures in the citizen–state relation, where the state's capacity to provide safety and security is brought under question. It is more of a generic terminology that indicates a variant of activities in cyberspace involving a number of different individuals, groups and organizations. To quote the National Research Council, "Tomorrow's terrorist may be able to do more damage with a keyboard than with a bomb" (NRC, 1991). Dorothy E. Denning, in her statement before the US Congress in 2000, defines cyberterrorism as:

> the convergence of terrorism and cyberspace. It is generally understood to mean unlawful attacks and threats of attack against computers, networks,

---

[2] Air-gap refers to "computers or networks that are not connected directly to the internet or to any other computers that are connected to the internet".

and the information stored therein when done to intimidate or coerce a government or its people in furtherance of political or social objectives ... result in violence against persons or property, or at least enough harm to generate fear. (Denning, 2007)

Whether or not a particular cyber-related attack can be considered under the category of cyberterrorism depends on the intent of the attacker and the source of origin of the attack. However, in order to categorize any cyber-related attack as cyberterrorism, the intended effect has to be serious, in terms of human and financial casualties, with greater intensity of fear and anxiety of terror, particularly among the citizens in the targeted country. Terrorists use cyberspace in facilitating traditional forms of terrorist activities that include bombings. They spread their motto, slogans and messages through websites, social media and look to radicalize the population and increase recruitment in achieving their purpose. The Internet has currently emerged as the biggest platform in both communicating and coordinating their action (Schwab, 2016).

In 1996, the Massachusetts-based Internet Service Provider (ISP) was disabled, while its record-keeping system was damaged by a hacker allegedly linked with the White Supremacist movement. In the end, the hacker posted: "As you have yet to see true electronic terrorism. This is a promise" (Tekwani, 2006). In 1998, the Sri Lankan embassies were swamped with hundreds if not thousands of emails a day for weeks by a hacker(s) associated with Liberation Tigers of Tamil Eelam (LTTE). The intelligence community considers this as "the first known attack by terrorists against" a country's computer network systems. In 1998, Khalid Ibrahim, a member of the militant "Indian separatist group Harkat-ul-Ansar", tried to buy Defense Information Systems Network Equipment Manager (DEM), non-classified military-networking program of the Department of Defense (DoD) from hackers. He reportedly obtained classified information of the US government and data from India's Bhabha Atomic Research Centre (BARC) (McKay, 1998).

In Iraq and Syria, the Internet has become a tool for insurgents, separatists and terrorist organizations to disseminate their ideology and plans, for recruitment as well as for coordinating attacks in other parts of the world. Terrorist organizations and groups like Islamic State, Al-Qaeda and Hezbollah are becoming increasingly adept in using the Internet for maintaining and expanding their operations by particularly developing skill-sets and access to information for conducting effective holocaustic cyber-related attacks on their targets (USDoJ, 2020). In the current scenario, there is nevertheless an increasing incidence where the terrorist organizations tend to use cyberspace for a variety of iniquitous activities, including the promotion of their ideology, for gaining political support and even for planning and coordinating attacks. The institutional authorities and agencies in different nations have authorized strengthening the security protocol for enhancing authority, monitoring cyberspace and email communications for any potent terrorist activities. The government agencies, civilian and military as well as the private sector are potentially vulnerable to cyberterrorism, and any attack on the CII can "have a multiplier effect on the" nation's security as a whole.

The third major dimension of threat in cyberspace is cybercrime. Cybercrime has emerged as one of the important dimensions of threats in (and from) the cyberspace, having a greater impact on the common netizens, even more often than the institutional actors. Cybercrime is a growing cause for concern for State as well as non-State actors. It could be considered as an extension of the traditional crime, but taking place in cyberspace. It is considered to be among the fastest-growing areas of crime, with several actors exploiting the speed, anonymity and accessibility for committing a wide array of criminal activities often transcend the "Westphalian" sovereign nature of physical borders causing significant harm as well as posing concrete but multiplier threats to victims anywhere in the world (Kreuder-Sonnen & Zangl, 2014).

Interpol loosely demarcates or compartmentalizes cybercrime into two, namely, *"advanced cybercrime"* and *"cyber-enabled crime"*. The former is considered high technology (high-tech) crime that initiates "sophisticated attacks against computer software and hardware". Meanwhile, the latter is considered to be evolved from traditional crimes due to the emergence and integration of the Internet within various societies. Meanwhile, a *cyber-enabled crime* includes the perpetuation of traditional and/or conventional forms of crime into cyberspace and primarily focuses on areas like financial crimes against children vis-à-vis., for pornography or pedophilic activities. Further, there were issues of credit card theft, vandalism, embezzlement, intellectual property theft, money laundering and even terrorism (INTERPOL, 2018, 5–18).

The prominent examples for *advanced cybercrime* include the Equifax Records Breach in September 2017, which revealed private information of more than 147 million US customers, including the "address, full name, social security number and even driver license number". It affected citizens in other countries related to the details of the accounts that were revealed. A compensation between 575 to 700 million USD was agreed to be paid by Equifax to compensate the victims of the attack (Colby, 2020).

Another form was the ransomware attacks in the '90s and the 2000s. It affected millions of people in hundreds of countries. Millions of dollars of money siphoned from victims blackmailed for access to their computer systems, online account details, personal information and other aspects. Until 2015, ransomware attacks averaged around four million ransomware attacks however saw an exponential rise leading to unmitigated disasters worldwide. In 2016 alone, there were around 638 million ransomware attacks, while the period between 2017–2019 averaged 192 million ransomware attacks (Statista, 2021a). As of 2019, the most prominent causes of ransomware infections were "spam/phishing emails" (67%), lack of cybersecurity training (36%), "weak passwords/access management" (30%), "poor user practices/gullibility" (25%), "malicious websites/web ads" (16%) and report clickbait (16%) (Statista, 2021b). It was seen that Windows Operating Systems (OS) amounted to 87 per cent of the total cases of ransomware attacks, which reveals huge loopholes being utilized by hackers (Statista, 2021c).

Cyberspace often offers more profit and less risk for various actors (stakeholders) in ensuing the sustenance of their criminal activities, and at many times without even a trace of evidence for prosecuting the criminals. Undoubtedly it often shatters the normative ideas of crime and justice, social values as well as ethics. The perpetrator of a crime does not see the victim, which is not even recognized. This skill of concealing the digital footprint of the perpetrator, as well as the ease of planting (fake) evidence is a real challenge for the delivery of justice and the protection of the innocent. Understanding the issues related to cybercrime is thus a prerequisite in dealing with the challenges it poses to society. This is possible through the analysis of the history and evolution of cyberthreats.

Thus, the cyber domain and the related spatial paradigm transcend any conventionally "perceived form of demarcated and delineated sovereign" political boundaries in the physical space–time paradigm. Globally, it is accepted as a public good and is generally used by netizens, by companies with commercial interests, by the State and non-State actors, particularly by military, mafia and other anti-State groupings, organizations, institutions and/or personnel. The victims of cyber-related attacks range from a wide range of sectors, including public health, banking, telecom, education, non-governmental actors, State actors or countries to international organizations and inter-governmental agencies. An example in this regard is that the critical national infrastructure (CNI) could also be affected by cyberthreats and inherently threaten the country's national security (Malik, 2004; McKinsey & Company, 2016).

The fourth dimension is *cyberwar* (also *cyberwarfare*). Cyberwar terminology is blurred in terms of definitions and examples, particularly in relation to the other three concepts: cybercrime, cyberespionage and cyberterrorism. The term cyberwarfare originated in the twenty-first century only (unlike the other three) though it has been in practice and existence at times before that (Orr, 2018; Puyvelde, 2015). However, it is yet to be fully understood, comprehended and/or agreed upon universally, except for differing definitions from different State actors and international organizations. This is often not just due to the transcending nature of actors and activities or the inaccuracy in demarcating or pinpointing the nature of the real source behind an attack but also due to differing perspectives and perception of different school of thought and narrative. It indicates an attack by a nation (or international organization) over another nation through cyberspace. It involves the actions as well as activities in enabling (to attack another nation's computers or information networks (RAND Corporation, 2020), often causing "significant death, damage or destruction" (Ranger, 2017).

The three major methods of cyberwarfare include sabotage (deliberately destroy, damage or obstruct), electronic espionage (steal information from computers using viruses) and finally, attacks on electrical (smart) power grids (Orr, 2018). Anders Fogh Rasmussen, the former NATO Secretary-General, once remarked cyberthreats "have become a new form of permanent, low-level warfare" (Wilson, 2011), requiring long-term security cooperation as well as collaboration to counter. Cyberwar (with the emergence of hybrid warfare) has often amplified the success of military campaigns through conventional warfare (Puyvelde, 2015). China is considered the first state actor to identify the potential of cyberwarfare and has established the world's largest contingent of dedicated offensive and defensive military units, with the intent to become a "cyber superpower" (Wilson, 2011).

In the twenty-first century, major state actors like China, the US, Russia, Iran, North Korea and Syria are prominently engaged in cyberwarfare. In 2006, there were reports of Chinese hacking into US

Air Force computers, servers and devices, US Military War College networks, National Aeronautics and Space Administration (NASA) during its launch, as well as the House of Commons computer system. Another prominent example of a cyber-related attack occurred in May 2007 against the Estonian government ministries, websites and critical infrastructure, which is considered a "modern form of hybrid warfare" (McGuinness, 2017). This was allegedly said to be originating from Russia as a reaction against the dismantling/relocation/removal of a Soviet-era Bronze Memorial Statue from its capital Tallinn in 2007. A large number of malicious viruses and trojans were sent to computers and devices connected and/or networked together and were collectively rendered inoperable to their original users. The information stored in these systems is transmitted and wiped clean of the respective system, leaving no digital footprints for further tracking (Davis, 2007).

There is an increasing number of emerging attacks in different and varying forms, developing new tools for intrusion. Countries like the US, China and possibly Russia are the current frontrunners in developing and integrating emergent technologies, especially AI, into their systems and capabilities (Bey, 2018, 31–35; Vincent, 2017). The next-gen cyberwarfare will be done by countries that have developed and enabled technologically advanced Artificial Intelligence (AI) into their cyber capabilities. Interestingly, "India has turned its attention" to developing AI in dealing with cyberwarfare, particularly with the growing cyber-related capabilities as well as attacks from its neighbours, China and Pakistan (Fay & Trenholm n.d.; Maymi & Lathrop, 2018, 71–80; Mi-hyun, 2018). This is intricately linked to the institutionalization of Digital India and Smart Cities initiatives in India (MeitY, 2021). Further, the emergence of high-speed interconnectivity and potential vulnerabilities from the integration of AI has been much sought after. The need to improve resilience, interoperability, integration of existing, emergent and future technologies has become a necessity (CISA, 2020; Cayamcela & Lim, 2018 TRAI, 2020; Smee & Hou, 2020). This is critical in dealing with risks and vulnerabilities in various sectors, especially supply chain, network security, deployment, competition and choice on the use of technologies (DHS, 2020b).

DIGITAL INDIA

India is an emerging power that is sometimes perceived to possibly threaten the rise (or status quo) of other major powers in the world in almost all domains of power. It is a major force in the land, sea (or ocean), air and outer space. Nevertheless, it needs to enforce its power in the cyber domain or cyberspace. The Indian position has been improving in terms of technical and technological know-how, several skilled personnel, collaboration and coordination between various agencies in the country, and its positive contribution of technological and technical advancements. However, along with this, there have arisen many cons to the emergence of cyberspace.

The former President of India, Dr. A.P.J. Abdul Kalam, who was also a physicist and known as the "Father of Indian Missile Technology" once opined that cyberwarfare "is the biggest threat to national security which will render even the Inter-Continental Ballistic Missiles (ICBMs) insignificant as a security threat" (Dilipraj & Reghunadhan, 2017, 115). There has been the rise of inimical actors and activities in cyberspace, with implications permeating into the physical world as well. The influence of State and non-State actors had a huge effect, especially due to physical–virtual interlinkages. This has often created issues of radicalization, terrorism, separatism, gender violence, identity theft, violence against children, privacy-related issues, threats to national security and many more (UNESCO, 2017).

The emerging challenges, vulnerabilities and threats on India's critical infrastructure, increasing attacks on the personnel, military and civilian infrastructure, individual users are all worrisome. It has had implications on the concept and understanding of territory and territorialization, sovereignty, application of law and economic overtures related to borders and boundaries in cyberspace. It has been the target of attacks by a different set of actors that have seen frequent data breaches, malware attacks and trojan intrusions, all of which have taken a toll on the strategic, government, private and civilian institutions (Swinhoe, 2021). The inherent nature of the cyberthreats in cyberspace, with the magnification and interconnectedness, coupled with the lack of coherent, coordinated institutional-cum-regulatory mechanisms, have made the country potentially vulnerable to various threats (Samuel & Sharma, 2016).

CHAPTER SCHEME

This first chapter provides an overall technological, societal and international politics of cyberspace, besides providing a brief overview of the dimensions of cyberspace. Chapter 2 deals with the global scenario of the threat landscape in cyberspace. This is an analysis based on the study of the historical overview of cyberthreats, well into the twenty-first century, and the impact on nation-states and their relations and actions. It also analyses the theoretical dimensions and international debates of cybersecurity, international relations and international political economy. Chapter 3 deals with the analysis of Digital India and the emerging threat landscape in cyberspace. In this chapter, an analysis of the nine pillars of Digital India is undertaken, further with an analysis of policy, organizational structure and governance are being critically analysed. Chapter 4 summarizes all the previous chapters while entailing on the overall structural and institutional framework, and provides policy recommendations and suggestions on improving cyber technological threat landscape in India. The book undertakes a study of the existing framework on the overall scenario of cyberthreats, the organizational structure, policing and diplomacy. An overview of cyberthreats and implications of cyberthreats in Digital India is dealt with. Several institutional visits were undertaken, and extensive interactions were made with experts and scholars in the field. A content analysis of various reports and literature from the international and national actors, agencies and organizations was also undertaken to provide analysis intended to bring out the inherent cyberthreats in India.

REFERENCES

Baines, S. (2011). Data espionage: A new age of spying. *Orange Business Services*. Retrieved March 2, 2019, from https://www.orange-business.com/en/blogs/connecting-technology/security/data-espionage-a-new-age-of-spying

Barlow, J. P. (1996). A declaration of the independence of cyberspace. *Electronic Frontier Foundation*. Retrieved May 2, 2018, from https://www.eff.org/cyberspace-independence

Bell, D. (2009). Cyberspace/Cyberculture. *International Encyclopedia of Human Geography*, 468–471.

Bey, M. (2018). Great powers in cyberspace: The strategic drivers behind US, Chinese and Russian competition. *The Cyber Defense Review, 3*(3), 31–37.

Brown, K. (2014). *The information umbrella.* Retrieved April 11, 2021, from https://aimblog.uoregon.edu/2014/04/15/our-shrinking-world/#.Wv2 BNHV96xU

Cayamcela, M. E. M., & Lim, W. (2018). Artificial intelligence in 5G technology: A survey. *2018 International Conference on Information and Communication Technology Convergence (ICTC),* 860–865

Cobb, S. (2014). Botnet malware: What it is and how to fight it. *WeLiveSecurity.* Retrieved December 10, 2018, from https://www.welivesecurity.com/2014/10/22/botnet-malware-fight/

Colby, C. (2020). You're running out of time to submit your Equifax data breach claim—Here's how. *Computer Network (CNET).* Retrieved January 11, 2021, from https://www.cnet.com/how-to/youre-running-out-of-time-to-submit-your-equifax-data-breach-claim-heres-how/

Consumer News and Business Channel (CNBC). (2012). *Investigations Inc.: Cyber Espionage.* Retrieved February 2, 2018, from https://www.cnbc.com/investigations-inc-cyber-espionage/

Cybersecurity & Infrastructure Security Agency (CISA). (2020). *CISA 5G STRATEGY: Ensuring the security and resilience of 5G infrastructure in our nation.* US Department of Homeland Security. Retrieved January 30, 2021, from https://www.cisa.gov/publication/5g-strategy

Davis, J. (2007). Hackers take down the most wired country in Europe. *WIRED.* Retrieved August 4, 2018, from https://www.wired.com/2007/08/ff-est onia/

Davis, N. (2016). What is the fourth industrial revolution?. *World Economic Forum.* Retrieved May 19, 2019, from https://www.weforum.org/agenda/2016/01/what-is-the-fourth-industrial-revolution/

Denning, D. E. (2007). Cyberterrorism—Testimony before the special oversight panel on terrorism committee on armed services U.S. house of representatives. In E. V. Linden (Ed.), *Focus on terrorism* (Vol. 9, pp. 71–76). Nova Science Publishers, Inc.

Department of Homeland Security (DHS). (2020a). *Cybersecurity.* Retrieved February 5, 2021, from https://www.dhs.gov/topic/cybersecurity

Department of Homeland Security (DHS). (2020b). *Feature article: 5G introduces new benefits, cybersecurity risks.* Retrieved February 7, 2021, from https://www.dhs.gov/science-and-technology/news/2020/10/15/fea ture-article-5g-introduces-new-benefits-cybersecurity-risks

Dilipraj, E., & Reghunadhan, R. (2017). Organisational governance of cyber space in India. *Journal of Air Power and Space Studies, 13*(1), 115–134.

Dufva, T., & Dufva, M. (2019). Grasping the future of the digital society. *Futures, 107,* 17–28.

Fay, R. & Trenholm, W. W. (n.d.). The cyber security battlefield: AI technology offers both opportunities and threats. *Centre for International Governance*

Innovation. Retrieved January 25, 2021, from https://www.cigionline.org/articles/cyber-security-battlefield

Featherly, K. (2021). Neuromancer. *Encyclopaedia britannica.* Retrieved February 7, 2021, from https://www.britannica.com/topic/Neuromancer

Gibson, W. (2004). *Neuromancer.* Penguin Random House LLC.

Goodreads. (2018). *The complete sprawl trilogy: Neuromancer, count zero, mona lisa overdrive.* Retrieved January 25, 2020, from https://www.goodreads.com/book/show/26891629-the-complete-sprawl-trilogy?rating=1&utm_medium=api&utm_source=blog_book

Hall, A. O. (2018). Cyber conflict in a competitive world. *The Cyber Defense Review, 3*(3), 9–14.

International Policing Body (INTERPOL). (2018). *Interpol: Specialized committee.* Retrieved October 17, 2020, from https://static1.squarespace.com/static/533e6b7de4b0d84a3bd7c4be/t/5a387def71c10b6d1365dabc/1513651697948/Interpol_Background_Guide.pdf, 1–25

Kirsch, S. (1995). The incredible shrinking world? technology and the production of space. *Environment and Planning d: Society and Society, 13*(5), 529–555.

Kreuder-Sonnen, C., & Zangl, B. (2014). Which post-Westphalia? International organizations between constitutionalism and authoritarianism. *European Journal of International Relations, 21*(3), 568–594.

Kumar, S. (2014). *Indias national security: Annual review 2013* (1st ed.). New Delhi: Routledge India.

Kushner, D. (2013). *The real story of stuxnet.* Retrieved February 12, 2020, from https://spectrum.ieee.org/telecom/security/the-real-story-of-stuxnet

Lagarde, C. (2018). Estimating cyber risk for the financial sector. *International monetary fund blog.* Retrieved May 19, 2020, from https://blogs.imf.org/2018/06/22/estimating-cyber-risk-for-the-financial-sector/

Laasme, H. (2011). *Estonia: Cyber window into the future of NATO.* Retrieved May 12, 2021, from https://www.thefreelibrary.com/Estonia%3A+cyber+window+into+the+future+of+NATO-a0275489814

Maher, K. (2013). *The new Westphalian web.* Retrieved May 22, 2021, from http://foreignpolicy.com/2013/02/25/the-new-westphalian-web/

Maizland, L. (2020). China's modernizing military. *Council for foreign relations.* Retrieved January 12, 2021, from https://www.cfr.org/backgrounder/chinas-modernizing-military

Malik, A. (2004). *Technology and Security in the 21st Century A Demand-side Perspective.* SIPRI Research Report No. 20. New York: Oxford University Press.

Maymi, F., & Lathrop, S. (2018). AI in cyberspace: Beyond the hype. *The Cyber Defense Review, 3*(3), 71–82.

McGuinness, D. (2017). How a cyber attack transformed Estonia. *BBC*. Retrieved December 8, 2019, from https://www.bbc.com/news/39655415

McKay, N. (1998). *Do terrorist troll the net?* Retrieved August 12, 2018, from https://www.wired.com/1998/11/do-terrorists-troll-the-net/

McKinsey & Company. (2016). *Digital globalization: The new era of global flows*. Retrieved March 13, 2018, from https://www.mckinsey.com/~/media/McKinsey/Business%20Functions/McKinsey%20Digital/Our%20Insights/Digital%20globalization%20The%20new%20era%20of%20global%20flows/MGI-Digital-globalization-Full-report.ashx, 1–156.

Mi-hyun, L. (2018). *Southeast Asia begins to prepare for cyber war; India turns to AI*. Retrieved February 2, 2018, from https://www.huffingtonpost.com/asiatoday/southeast-asia-begins-to_b_14334812.html

Ministry of Electronics & Information Technology (MeitY). (2019). *Report of committee—A on platforms and data on Artificial Intelligence*. Retrieved December 1, 2020, from https://www.meity.gov.in/artificial-intelligence-committees-reports

Ministry of Electronics & Information Technology (MeitY). (2021). *Smart cities mission*. Retrieved February 7, 2021, from https://indiaai.gov.in/missions/smart-cities-mission

National Research Council (NRC). (1991). *Computers at risk: Safe computing in the information age*. The National Academies Press.

O'Brien, D. (2017). *A short history of cyber espionage*. Retrieved March 12, 2019, from https://medium.com/threat-intel/cyber-espionage-spying-409416c794ec

Orr, T. (2018). A brief history of cyber warfare. *GRA Quantum*. Retrieved May 2, 2019, from https://graquantum.com/a-brief-history-of-cyberwarfare/

Puyvelde, D. V. (2015). Hybrid war—Does it even exist?. *Nato Review*. Retrieved February 2, 2020, from https://www.nato.int/docu/review/2015/Also-in-2015/hybrid-modern-future-warfare-russia-ukraine/EN/

RAND Corporation. (2020). *Cyber warfare*. Retrieved January 2, 2021, from https://www.rand.org/topics/cyber-warfare.html

Ranger, S. (2017). *Cyberwar: A guide to the frightening future of online conflict*. Retrieved June 2, 2020, from http://www.zdnet.com/article/cyberwar-a-guide-to-the-frightening-future-of-online-conflict/

Reardon, R. & Nazli, C. (2012). *The role of cyberspace in international relations: A view of the literature*. Retrieved July 15, 2021, from https://ecir.mit.edu/sites/default/files/documents/%5BReardon%2C%20Choucri%5D%202012%20The%20Role%20of%20Cyberspace%20in%20International%20Relations.pdf

Rubenstein, D. (2014). *Nation state cyber espionage and its impacts*. Retrieved July 9, 2019, from https://www.cse.wustl.edu/~jain/cse571-14/ftp/cyber_espionage/

Samuel, C., & Sharma, M. (2016). *Securing cyberspace: International and Asian perspectives*. Pentagon Press.

Schwab, K. (2016). *The fourth industrial revolution*. World Economic Forum.

Schweitzer, Y., Siboni, G., & Yogev, E. (2013). Cyberspace and terrorist organizations. In G. Siboni (Ed.), *Cyberspace and national security: Selected articles*. Tel Aviv.

Smee, J. E., & Hou, J. (2020). 5G+AI: The ingredients fueling tomorrow's tech innovations. *Qualcomm*. Retrieved August 10, 2020, from https://www.qualcomm.com/news/onq/2020/02/04/5gai-ingredients-fueling-tomorrows-tech-innovations

Statista. (2021a). *Annual number of ransomware attacks worldwide from 2014 to 2019*. Retrieved February 3, 2021, from https://www.statista.com/statistics/494947/ransomware-attacks-per-year-worldwide/

Statista. (2021b). *Most common delivery methods and cybersecurity vulnerabilities causing ransomware infections according to MSPs worldwide as of 2019*. Retrieved February 3, 2021, from https://www.statista.com/statistics/700965/leading-cause-of-ransomware-infection/

Statista. (2021c). *What systems have you seen infected by ransomware?* Retrieved February 3, 2021, from https://www.statista.com/statistics/701020/major-operating-systems-targeted-by-ransomware/

Stockholm International Peace Research Institute (SIPRI). (2020). *SIPRI yearbook 2020: Armaments, disarmament and international security*. SIPRI.

Swinhoe, D. (2021). The 15 biggest data breaches of the 21st century. *CSO India*. Retrieved February 5, 2021, from https://www.csoonline.com/article/2130877/the-biggest-data-breaches-of-the-21st-century.html

Tekwani, S. (2006). *The LTTE's online network and its implications for regional security*. Working Paper. Retrieved June 20, 2019, from https://www.rsis.edu.sg/wp-content/uploads/rsis-pubs/WP104.pdf

Telecom Regulatory Authority of India (TRAI). (2020). *A white paper on smart cities in India: Framework for ICT infrastructure*. Mahanagar Doorsanchar Bhawan.

The Economist. (2010). *War in the fifth domain*. Retrieved April 10, 2021, from http://www.economist.com/node/16478792

United Nations Educational, Scientific and Cultural Organization (UNESCO). (2017). Youth and violent extremism on social media: Mapping the research. Retrieved December 2019, from https://unesdoc.unesco.org/ark:/48223/pf0000260382

United States Department of Justice (USDoJ). (2020). Global disruption of three terror finance cyber-enabled campaigns: Largest ever seizure of terrorist organizations' cryptocurrency accounts. *Office of public affairs*. Retrieved January 19, 2021, from https://www.justice.gov/opa/pr/global-disruption-three-terror-finance-cyber-enabled-campaigns

Vincent, J. (2017). *Putin says the nation that leads in AI 'will be the ruler of the world'*. Retrieved February 2, 2018, from https://www.theverge.com/2017/9/4/16251226/russia-ai-putin-rule-the-world

Wilson, J. R. (2011). *A state of permanent warfare?* Retrieved February 2, 2018, from https://www.defensemedianetwork.com/stories/a-state-of-permanent-warfare/

Zetter, K. (2012). *Meet 'Flame,' the massive spy malware infiltrating Iranian computers*. Retrieved April 2, 2018, from https://www.wired.com/2012/05/flame/

Zetter, K. (2014). *Hacker Lexicon: What is an air gap?* Retrieved April 2, 2018, from https://www.wired.com/2014/12/hacker-lexicon-air-gap/

History and Evolution of Global Cyber Technological Threat Landscape: Theoretical Dimensions

Abstract The chapter entails the history and evolution of various cyber technologies across the globe, and focuses on the particular development of the threat landscape in the cyberspace and its perpetuation into the physical world. Further, the development of academic and scholarly understanding of the area, especially the theoretical dimensions have been dealt with in detail.

Keywords History · Global evolution · Cyber threat landscape · Theoretical dimensions · International political economy

Introduction: Early Period (1960s–1990s)

In 2020, the 46th anniversary of the Denial of Service (DoS) attack was remembered. It was for the first time that an attack on computer systems was recorded in the 1960s. Being born as the brainchild of a 13-year-old high school student David Dennis, this is considered the first form of cyber threat. The attack was directed on "the Programmed Logic for Automatic Teaching Operations (PLATO)" terminals; the first computerized shared learning systems developed by the "Computer-Based Education Research Laboratory (CERL)" at the University of Illinois at Urbana Champaign (UIUC) in 1960. PLATO was developed

© The Author(s), under exclusive license to Springer Nature Singapore Pte Ltd. 2022
R. Reghunadhan, *Cyber Technological Paradigms and Threat Landscape in India*, https://doi.org/10.1007/978-981-16-9128-7_2

by Donald L. Bitzer, which is considered the precursor of many current and future multi-user computer systems (Dear, 2017; *PLATO History Foundation*, 2010).

The year 1971 was an important year for the emergence and revolution of technology of two kinds, particularly for the emergence of technological paradigms in cyberspace. In 1971, the *microchip* was invented, which led to the emergence, evolution and exponential transformation of emergent/frontier technological paradigms. This included robotics, artificial intelligence (AI), etc., all of which was (and is) interlinked to the socio-technological metamorphosis and transmutation of individuals, societies, nations across the globe (Cottam, 2020). The same year, an experimental worm called *Creeper*, the world's first self-replicating program developed for the first time by Robert (Bob) H. Thomas and his team at BBN Technologies (originally Bolt, Beranek and Newman). The purpose of the worm was to check for loopholes and inefficiencies (as in *whitehat* activities) within the computer systems developed and tested. The worm was used to infect the computers running the *TENEX* operating system. This period also saw the development and use of first email programs,[1] the first use of "@" symbol in email addresses. The period also saw the use of "Advanced Research Projects Agency Network (ARPANET)", the predecessor of the modern-day internet.[2] John von Neumann, a US-based mathematician, is said to have "envisioned specialized computers or *self-replicating automata*,[3] that could build copies of themselves and pass on their programming to their progeny" (Dalakov, 2018; Spafford, 1989, 17–57). This paved way for the development of the first antivirus software by BBN technologies, known as *Reaper*. It was developed as part of ethical hacking and vulnerability testing of their system(s) and software(s).

Within the next decade, the world saw the "first large-scale computer virus" outbreak in history, called *Elk Cloner*. This was followed by the *Wabbit virus* (in 1974) and *Pervading Animal* (in 1974–75). In 1982, a 15-year-old named Richard Skrenta wrote a computer virus for the

---

[1] Like SNDMSG and READMAIL. Both were local inter-user mail programs. It was written by Ray Tomlinson of BBN Technologies.

[2] Though used widely in the early 1970s, its operation originally began in the late 1960s.

[3] The self-replicating program, called *virus* or *computer virus* (first of its kind), could self-replicate over a computer network or internet and was known as a worm.

"Apple-II operating system" through floppy diskettes, which forcefully booted the computer (*Techopedia*, 2018b). This led to the development of what became known as the *boot sector virus* in the latter half of the 1980s. Interestingly, the term *virus* was only coined in 1983 by Frederick Cohen, and only in 1984 did he use the term *computer virus* (on the suggestion of his teacher Leonard Adleman). The *Brain boot sector* virus (also called *Brain virus* or *Pakistani Flu*) was first detected virus (in 1986) in personal computer platforms that predominantly ran on "PC-DOS (Personal Computer – Disk Operating System)" and/or IBM-DOS. It was infected through floppy disks, replicating in a manner, so as to boot the computer completely and had greater stealth qualities (Hipponen, 2011).

The *Cuckoo's Egg* (in 1985–87), one which is considered to be malware used by German hackers under KGB, precedes the *Brain boot sector* virus attack. These hackers developed under the West German Intelligence Agency (BND), focusing on cyberespionage under Project RAHAB, which began originally in 1981 until (officially) it was shut down in 1991 (AFCEA, 2018). In 1987, the IBM platform was infected for the first time by the *Vienna virus*. This was followed by a series of viruses in that year, namely *Lehigh virus*, *Jerusalem virus*, *Swiss Cracking Association* (*SCA virus*), *Christmas Tree EXEC*. This paralysed "a large number of computer systems" and connected devices around the globe (Brookes, 2011; Cluley, 2010; *Techopedia*, 2018c; van Wyk, 1989). In June 1988, the *Festering Hate Apple ProDOS virus* was created by a group identified as the "Legion of Doom" that contained the notoriously famous members like Patrick Karel Koupa (a.k.a *Lord Digital*) and Bruce Fancher (a.k.a *Dead Lord*) typically targeting the Apple OSs and often destroying the computer(s) in the process (Weyhrich, 2010).

In November 1988, the *Morris* worm developed by Robert Tappan Morris became more notoriously famous in the history of cyber attack. A computer graduate student "at the Massachusetts Institute of Technology (MIT)", Morris released the worm at Cornell University, particularly through internet-connected computer devices that ran on a variant of UNIX OS systems. It was a self-replicating worm that slowed down the computer systems but was eventually detected by the authorities. He became "the first individual to be tried under the new" Act, called the *Computer Fraud and Abuse Act* of 1986, specifically created for the purpose of containing and restricting computer-related fraudulent activities and abuses. This later led to his conviction, the first of its kind in the history of humanity. This case led to "the creation of the Defense Advanced Research Projects Agency (DARPA)" for coordination concerning information sharing and institutionalization of response mechanism to any existing computer vulnerabilities and security. This case led to the development of the concept of "systemic risk" and emphasized the problems related to "exploitable vulnerabilities" for vendors and systemic administrators (Kelty, 2011; Shoch & Hupp, 1982, 172–180). However, interestingly, later on, Morris became a full-time Professor at MIT and taught there (Marsan, 2008).

In 1989, viruses like *BURP* and *Load Runner* created havoc in Apple II Oss (Weyhrich, 2010). Again in 1989, the "first ransomware virus called *AIDS Trojan*" was developed by Joseph L. Popp. However, interestingly Popp was an evolutionary biologist with a PhD from Harvard and was able to infect different users through the floppy diskettes, which he was able to distribute to numerous scientists and researchers in around 90 countries in the name of a survey that he supposedly conducted in understanding AIDS; a disease that was at its peak in infecting the population. Now ransomware can be defined as "a form of malicious software (in short called *malware*) that… [takes] over your computer, threatens [to harm] you… usually by [denial of] access to your data… [demanding] a ransom… to restore access to the data… [costing] few hundred dollars to thousands" (Fruhlinger, 2017b). INTERPOL defines ransomware as a kind of "malware [that denies access to a computer or mobile device, [particularly by encrypting] the data on a system, [and] demanding money for [the] restoration of the [functionalities]" (INTERPOL, 2020a, 2020b). This virus was created when the AIDS epidemic was at its peak and asked for a "licensing fee" ranging from 189 to 378 USD to be transferred to an address in Panama (Francis, 2016; Waddel, 2016).

In 1989–90, Mark Washburn, and Ralf Burger, developed the Chameleon family, which led to the development of the first computer virus in polymorphic encryption called *1260* or *V2PX* that specifically attacked and infected ".*com*" files, in turn affecting almost all files in a system (*Eyerys*, 1990). In March 1992, *Michelangelo* (a boot sector virus) was developed, spread through floppy diskettes and/or hard disks, and had a large-scale reach. Being developed in the pre-internet era, this was ostensibly the first malware that created a public scare at such a massive scale, with around five million devices being infected. Though the creator of *Michelangelo* was never found, this incident created greater awareness for computer and internet security, and increased the demand for antivirus and other security software as well as products (*TrendMicro*, 2017).

In 1993, Intel released its first Pentium processor systems, boot sector viruses *Leandro & Kelly* and *Freddy Krueger* spread quickly, utilizing the loopholes in its shareware distribution (Benson, 2014, 115). In 1994, Vladimir Leonidovich Levin, a Russian software engineer from St. Petersburg (in Russia), embezzled around 12 million USD from Citibank in New York (in the US) by hacking accounts and transferred money out of the bank's accounts; but was only able to withdraw 400,000 USD as the related accounts were frozen. Levin was caught along with other six accomplices and "was sentenced to three years in a US prison" along with a fine to pay. This increased focus for network security and other related measures to strengthen system security (Harmon, 1998; *WIRED*, 1998).

In 1995, the *Concept virus* came out, specifically targeting Microsoft Word documents but capable of attacking any OSs like IBM or Macintosh (Flashing Cursor). In 1996 the *Ply virus*, a 16-bit based polymorphic virus, started infecting systems. This further enabled the development and release of the much-advanced 32-bit viruses like the *ACG* virus in 1998 (Szor, 2005). In 1998 the first version of the *CIH* virus (also called the *Chernobyl virus*) came out and led to the Chernobyl nuclear power plant explosion, the largest nuclear accident globally. It became common and prevalent during the late 90 s, causing millions of dollars of damages to different nations. The virus is believed to have been created by Chen Ing Hau of Taipei's Tatung Institute of Technology in Taiwan was primarily intended for destruction; and thus, the virus getting its name *CIH v1.2 TTIT* or *CIH virus*. In 1999, *Happy99 worm* (January), the *Melissa worm* (March), *ExploreZip worm* (June) and *Sub7 Trojan horse* (December) were released with common features and similarities in attacks; particularly concentrating on the Windows OS users and its software programs like Microsoft Office and Outlook Express and was even able to steal passwords and credit card information (Bradley, 2016; Cluley, 2010; Chen, 2003; CMU, 1999).

In 1996, it was found that Paul Frederick Laney and two other men created the much-hyped Internet chat room called "KidsSexPics", and later "Orchid Club". This was established to engage in activities of exchanging details of child pornography through the Internet. The digital files contained pornographic photographs as well as videos of underage girls and boys and were distributed based on the demand put forth by the members of the group. The perpetrators were later caught and convicted, later leading to the passage of "the Child Pornography Protection Act in 1996" in the US, and the incorporation of technology-specific language. "[The US] Congress expanded the definition of child pornography to include altered pictures of... children... depictions ... of minors... in sexually explicit situations" (Britz, 2012, 193).

In the '90s, along with the popularization of Internet Relay Chat (IRC), cyber attacks saw an exponential rise. IRC is an application layer protocol that enables communication in textual form (chatting). These attacks primarily focused on gaining administrative privileges for users and in controlling IRC Channel. In the middle of 1999, groups used a tool known as *Trinoo* to "launch Distributed-Denial-of-Services (DDoS) attacks" with over 2000 infected systems all over the world. In August 1999, a much-advanced version of *Trinoo*, the *Tribe Flood Network* (TFN), was used to disable the "computer network of the University of Minnesota" for a few days. The "network of compromised machines", known as "Masters and Daemons", allowed the attacker(s) to "send a DoS instruction" to the Masters (main systems controlled by the hackers) and forward it to a large number of Daemons (zombie computers). Unfortunately, most of the victims had no idea of the scale of the attack or the loss of control and data from the affected computers (Dittrich, 1999; Jones, 2015; Howard, 2010; *Radware*, 2017; *Techopedia*, 2018a).

In the latter half of 1999, a TFN agent called *Stacheldraht* (called barbed wire in German) was used to attack and collect information from victims using Linux and Solaris operating systems and were mostly concentrated in the regions of the US and Europe. The method used was called *spoofing*, where communication was sent from a hidden and/or unknown source that is often disguised as a source with which the receiver is usually acquainted. It is more prevalent in devices, systems as well as communication mechanisms that lack resilient security measures. Later *Trinity* and *TFN2K* were developed, and newer tools like *mstream*, *Shaft* and *Omega* were commonly used by different hackers (Ibid.). In the latter half of 1999, the *WM97/Melissa-AG virus* (also called *Prilissa*) infected Windows Oss, mainly by infecting the documents and attaching them through email (Cluley, 2010).

CYBER THREAT LANDSCAPE IN THE TWENTY-FIRST CENTURY (2000S TO PRESENT)

The beginning of the twenty-first century saw an increasing number of attacks that were more motivated by the public attention it received and the attrition-oriented attribute it showed. The governments, companies, financial institutions were brought down by DDoS attacks. Between 2000 and 2002, nature, scope and the type of cyber attacks, grew in complexity, often influencing the geopolitics of different regions and economic wherewithal of various State(s) as well as non-State actor(s). The era saw a huge perpetuation of data exfiltration, which saw an exponential trend and widespread impact (Ullah et al., 2018, 18–54).

In the early period of the year 2000, a Russian hacker known as *"Maxus"* reportedly stole three lakh credit card numbers from the database of a music retailer, CD Universe; and created the largest credit card heist the world had ever seen. The account was traced to a bank in Hansabanka in Latvia under the name of Maxim Ivancov. These credit cards were sold by *Maxus* directly or through intermediaries. The card was used for different purposes by the end-users: online shopping, accessing child pornographic websites, possibly for money laundering, drug trafficking, buying weapons and/or even terrorism (Lyons, 2000). In the same year (in 2000), between March and August, two nationals of Kazakhstan, Oleg Zezov and Igor Yarimaka, were arrested for breaking into computer systems of the financial information company Bloomberg (in the New York office) and for blackmailing as well as extorting money in return for stolen passwords and records of the company and its personnel (Sullivan, 2000). The *W32/ Navidad virus* (in the latter period of 2000) and *Maldal virus* (in the early period of 2001) started infecting systems through emails, particularly through e-Christmas cards directly sent to the users (Cluley, 2010).

In January 2003, the *SQL Slammer worm* (also called *Sapphire worm*), first discovered by security analyst David Litchfield, became the fastest spreading worm and crashed the Internet within 15 min of its release (through DoS attack) (Kushner, 2013). Originally developed to test bugs in Microsoft SQL Server in 2002, mainly aspects related to bypassing "the prevention mechanisms", it infected more than 250,000 machines (30% of which were infected in a matter of minutes) and in turn affected nearly 27 million people. It led to a slowdown in network performance. The DoS attacks knocked out "internet and cell phone coverage,... and

ATMs" (Editor, 2016; *Techopedia*, 2020). This was similar to how Blaster Worm targeted Microsoft platforms as well (Schiller et al., 2007, 1–27). This opened the new arena for multiple developments and deployments of worms, viruses and related attacks worldwide. In the same year, the *Blackout virus* (also called *Apocalypse I* or *Apocalypse II*) affected the Apple II OSs through 3.5' disks, specifically targeting the security features (Weyhrich, 2010).

In 2004, the *Zafi-D virus* created in Hungary, infected users by sending seasonal greetings attached to their emails (Cluley, 2010). In 2005–05, more than a million computers and devices were infected from *Storm malware*, reportedly was created by Zhelatin Group, a Russia-based hacker group (Garretson, 2007). In comparison to malware like *SQL Slammer* and *Blaster*, it was less destructive and occasional in its DoS attacks. In 2006, the *Archiveus Trojan*, which uses a new form of the *Zafi-D virus*, was released. Its password protects files and asks the users to buy a product online in exchange for the password (Francis, 2016; Hayashi, 2010). In the same year, the *Oompa-Loompa virus* (also called *Leap*) became the first virus to infect Apple X OS, famously exposing the (potential) vulnerabilities in it (Brookes, 2011). *Operation Shady RAT* (2006–11) was one of the major hacking events that saw infiltration of "the computer systems of national governments, global corporations, nonprofits, and other organizations, with more than 70 victims in 14 countries" (Gross, 2011).

In January 2007, millions of systems were infected through a DDoS attack by the *Storm Worm Trojan*, specifically targeting Windows users to download malware into their devices and/ or systems using spams, links to online videos and even greeting cards. It is predicted to have accumulated the speed of a supercomputer, capable of bringing down almost anything from a government to a major corporation within a few minutes (Naraine, 2007; Stewart, 2007). In July 2007, the *Zeus Trojan* malware infected computers and stole a large amount of data, replicating itself in the process through instant messaging as well as spam emails. It could be considered as one of the most effective and successful malware programs of its time, having been developed into more than 50,000 variants during this period (Beal, 2018). There were issues with copyright infringement and/or information theft reaching massive proportions through popular file-sharing programs like Morphous, Limewire, Napster and Kazaa. The copyrighted songs were illegally sold in millions, costing the original owners billions of dollars (*Kaspersky*, 2018; Moore, 2015).

In August 2007, countries like Germany, the UK and France complained of Chinese intrusion into their networks. This strained relations between China and the three European powers, often leading to a war of words between politicians of respective nations through public speeches. In September 2007, Israeli forces hacked into Syrian Air Defense Networks, to allegedly enable bombing and disabling of Syrian nuclear facility. In October 2007, China alleged that they were attacked by hackers prominently from Taiwan and the US. In 2008, Georgia was attacked supposedly by hackers coordinated by the Russian military, defacing government websites, (CSIS, 2021) though some claim that "nationalist Russian hackers acted on their own" (Shactman, 2008). In 2008, a worm called *Conflicker* (also known as *Downadup*) infected millions of computers and devices, particularly targeting a vulnerability in Windows Operating System (OS). This enabled cybercriminals for spamming, phishing, conducting identity theft and other malicious activities, extended into 2009 (Cobb, 2014; *TechTarget*, 2009).

In 2009–10, "*Stuxnet* was discovered by a Belarus-based security firm", which led to the altering of programmable logic controllers (PLCs) in nuclear plants. This is considered a category of "Windows zero-day vulnerabilities", the first of its kind that repeatedly targeted Iran's nuclear programme and was reportedly initiated by the US by George Bush Jr and the Obama Administration under the code name *Operation Olympic Games* (Fruhlinger, 2017c). Social media targeting by malware programs became prominent during this period. The *Koobface worm* came to the fore by targeting netizens in social media like Facebook; even used a website theme of Santa to trap users (Cluley, 2010). This was used for cyber propaganda, propagating fake news to form psychological attacks on the personnel and general citizens. The latest example is Cambridge Analytica's case, who purportedly uses the online user data from Facebook to "persuade" (manipulate) the decision-making by voters in various countries (Meredith, 2018). Additionally, the *GhostNet* (2009) incident emerged, where a series of cyberespionage activities reportedly compromised more than thousands of devices in hundreds of countries, all of which originated in the Chinese mainland (Cluley, 2009).

The year 2011 was dubbed as the "Year of Hack", (Gross, 2011) where about five million computers in the US were infected by a sophisticated malware called *TDSS* (also known as *Alureon* or *TDL-4*). It attacked through spamming, malware downloading, DoS attacks, identity theft as well as different types of online fraud (Rouse, 2011). In

2011, a variant (and probable precursor to the *Stuxnet* malware), known as *Duqu*, was discovered by US-based security provider Symantec to have infected devices mostly in Europe. The malware was "designed to conduct reconnaissance on an unknown industrial control system and gather intelligence that can later be used to conduct a targeted attack" (Johnson, 2011; Zetter, 2011). Simultaneously, a malicious software program called *Flashback Trojan* infected around four to five million Macbooks and was discovered by Russian antivirus vendor Intego (Mathis, 2011).

Later, an unnamed Trojan that mimicked the Windows Product Activation Code came to the fore. It infected computer systems, extorting huge charges from different victims while offering to connect to an international number that it claimed was free. In 2012, a ransomware Trojan named *Reveton* spread to Europe and extorted cash by making the victim pay using a voucher from a cash service kept anonymous. This later led to further development of "police-based" ransomware in the likes of *Urausy* and *Tohfy*. However, the most important incident of the year was the cyberespionage incident involving the *Flame* malware, "discovered by the Russia-based Kaspersky Lab". It targeted and infected devices and "systems in Iran, Lebanon, Syria, Sudan, the Israeli Occupied Territories and other countries in the Middle East and North Africa" (Zetter, 2010, 2012). This was followed by the *Gauss* Trojan state cyberespionage programme, originally found during an investigation by the International Telecommunications Union (ITU) (*Kaspersky*, 2012).

Further, the attack by the *Shamoon* virus targeting the CNI in Saudi Arabia led to a "sudden disruption" of the oil and energy sector in the country (BBC, 2021). In January 2012, a banking malware *Cridex* was discovered, stole banking information, "credentials and other personal information to gain access" to a user's financial data. It contained features of *GameOver Zeus* malware, which led to the development of variants like *Bugat* and *Feodo* (Stroud 2018a). *NetWire* is "a remote access tool (RAT)" malware that led to widespread cyber attacks through identity thefts, further affecting the start-up menu of Windows OS systems (Meskauskas, 2021; NJCCIC, 2016). Interestingly, a variant of the *Stuxnet*, known as *Havex* malware came out in 2013. It targeted organizations mainly in the "US, EU and Canada" (*McAfee*, 2021).

Between July–September 2013, South Korean institutions, think tanks, government and defence industries were reportedly attacked and rendered inoperative by North Korean hackers in many cases. In September 2013, Iran reportedly hacked into unclassified devices and computers of the US

Navy. In September 2013, the first cryptographic malware called *CryptoLocker* was created. It compromised websites, phished and/or spammed business personnel, mimicking consumer complaints, and prominently used the *GameOver Zeus* botnet infrastructure. In 2014, ransomware using Tor as well as Bitcoin, called *CryptoDefense*, was released. It targeted Windows OS and later led to the development of much more aggressive *CryptoWall*. This included email spamming, identity theft and targeting unique ids accumulating an estimated 325 million USD. In the same year, the world's first android-based ransomware called *Sypeng* locked the screens of infected devices displaying FBI warning and delivered fake SMSs and Adobe Flash updates.

In April 2014, there emerged *Koler ransomware* that used "fake police penalties" and demanded ransom, and can be considered the first "*Lockerworm*"; due to its feature of sending messages and URLs to all contacts in the phone without the knowledge of the user and locking them out of their phone. A banking malware in the name of *Dridex* Trojan (that which evolved from *Cridex malware* of 2012) started spreading and infecting in 2014. It relied on malspamming (sending bulk spam emails), with an estimated 15,000 spam emails being sent each day and was primarily concentrated in the UK. It targeted Microsoft Office and used a zero-day vulnerability to infect Microsoft Word (*Computer Hope*, 2017; Kaimba, 2017; Stroud, 2018a; Stroud, 2018b). The *CTB-Locker* targeting Windows OS emerged in the same year, had multi-tier infrastructure, and used proxy machines and/or devices as botnets and depended on multiple Bitcoin (cryptocurrency) wallets (Sponchioni & Zhicheng, 2017).

In November 2014, "Sony Pictures Entertainment was hacked by a group" known as "the Guardians of Peace", allegedly based in North Korea (VanDerWerff & Lee, 2015). In 2014, the *SimpLocker*, which was the first "crypto-based ransomware", also named *Android.Simplocker* compromised the android devices of the user(s) and encrypted all the files in them. In turn, the users have to pay in Bitcoin or other forms of cryptocurrencies to get the device decrypted. In a majority of the cases the was not kept, and further demands for payment were done by the hackers (Sponchioni & Zhicheng, 2017). In 2015, the period between March and December targeted cyber attacks were undertaken on the Industrial Control Systems (ICS) of Ukraine's electrical grid. This was done using the *BlackEnergy 3* malware that created "a communication channel to the… command and control" of multiple "distribution *oblenergos*",[4] and thus created a "power outage" for 225,000 customers (Shehod, 2016, 1–5). Later, parts of the power grid were destroyed by the *KillDisk* malware (Cerulus, 2019).

Further, many variant forms of malwares and ransomwares emerged, like the *Lockerpin* (in 2015) "*Ransomware-as-a-Service* or *RaaS*" (in 2015), *TeslaCrypt* (in 2015), *LowLevel04* (in 2015), *Chimera* (in 2015), *Ransom32* (in 2016), *7ev3n* (in 2016), *LOcky* (in 2016), *SamSam* (in 2016), *KeRanger* (in 2016), *Petya* (in 2016), *Maktub* (in 2016), *Jigsaw* (in 2016), *CryptXXX* (in 2016), *ZCryptor* (in 2016), *Industroyer* (in 2016), *Triton* (in 2017) and *NotPetya* (2017) were all advanced versions of ransomware and malware programs, mostly upgraded, using multiple strains, multi-tier infrastructure and features targeting different systems and creating millions of victims in the process (Francis, 2016; Francis, 2016; Fruhlinger, 2017a; Hayashi, 2010; Knowbe4, 2018; *McAfee*, 2021). The use of *NotPetya* has been prominently seen in the activities of various Russian and affiliated hacking groups like "Fancy Bear, Cozy Bear and Sandworm" (Cerulus, 2019).

The Brexit referendum exposed potential and alleged DDoS cyber attacks in 2016, which often caused it to crash, affecting the voting process. In June 2016, around 25,000 "Digital Video Recorders (DVRs) and Closed-Circuit-Television (CCTV) cameras" were hacked, websites brought down with ten thousand HTTP requests received per second (Daniel, 2016). The prominent case during this year is the 2016 US elections, wherein it was found that Cambridge Analytica, a UK-based company appropriated the data (personally identifiable information or PII) of nearly one-fourth of Facebook users in the US. The PII was collected based on "their 'OCEAN' psychological profile (openness, conscientiousness, extraversion, agreeableness, and neuroticism) and correlated it with their Facebook activity (likes and shares)" (González, 2017, 9–12; Isaak & Hanna, 2018, 56–59).

Kaspersky reported that the banking sector was affected by malware families that were dominated by "*Trojan-Downloader.Win32.Upatre*" (42.36%), "*Trojan-Spy.Win32.Zbot*" (26.38%), "*Trojan-Banker.Win32.ChePro*" (9.22%), "*Trojan-Banker.Win32.Shiotob*" (5.1%), and "*Trojan-Banker.Win32.Banbra*"

---

[4] *Oblenergos* is a Ukrainian term for energy companies.

(3.51%). In terms of target nations, Singapore (11.6%), Austria (10.6%), Switzerland (10.6%), Australia (10.1%), and New Zealand (10%) were the top five affected countries. *Malicious URL* accounted for 75.76% of the total online attacks on a computer, with Russia (48.9%), Kazakhstan (46.27%), Azerbaijan (43.23%), Ukraine (40.4%), and Vietnam (39.55%) were the top five countries affected by online infections. The most targeted application by cyber attacks included *Browsers* (62%), *Android* (14%), *Java* (13%), *Adobe Flash Player* (four per cent), *Office* (four per cent) and *Adobe Reader* (three per cent) (Garnaeva et al., 2015).

In September 2016, similar to 2015, ransomware programs were used for two DDoS attacks of the scale of 665 Gbps (Krebbs, 2016) and one Tbps; against the website of Brian Krebbs and the OVH (a company based in France), respectively, the largest attack in the history of the Internet (*OVH*, 2016). In October 2016, DYN Inc (a US-based firm) was attacked using the same botnet that attacked "krebbsonsecurity.com" and "OVH" the previous month. Unlike Krebbs and the OVH, DYN Inc. could not withstand the attacks and went offline, affecting companies like Netflix, Amazon, Twitter and CNN (*Radware*, 2016). Later, it emerged that the botnet used to attack these Internet of Things (IoT) devices and related computers was the *DNS Waterfall Torture* (famously called *Mirai* botnet). It has an open-source code, meaning hackers anywhere in the world could access, use, alter and/or customize it based on the requirements and targets they have in mind (Jones, 2015). This has serious implications in terms of security aspects for various devices as well as the magnitude of each attack, particularly for a relatively low cost.

In January 2017, the "Swedish Foreign Policy Institute" alleged that Russia is conducting warfare to weaken it internally, while Pakistan reportedly undertook attacks on India (February); China on South Korea, the US, Europe and Japan (April); North Korea on the US (April–October); Israel on Lebanon (May), Iran on Britain (June), Russia on Ukraine and Montenegro (June); United Arab Emirates (UAE) on Qatar (July); North Korea on South Korea (August–October); the US on North Korea (September); Russia on NATO (September); China on the US (October); Russia on Poland (October); Russia on the US (October); Vietnam on ASEAN organizations (November).

Over the last few years, the US Director for National Intelligence and the Department of Homeland Security "jointly identified Russia as the responsible actor for hacking Democratic National Committee" and dumping emails from Wikileaks (CSIS, 2021). In February 2017, serious breaches were found in Singapore's military and related systems, while the "Central Bureau of Investigation (CBI)" in India has been reportedly targeted by phishing campaigns originating in Pakistan. In the middle of 2017, ransomware attacks were prevalent under the *WannaCry* ransomware (also named *Ransom.Wannacry* or *wcry*). The new strain of ransomware attacked more than 100 countries, infecting lakhs of systems and devices, including hospitals, SCADA systems in dams and even high-security military establishments. It was claimed to be done by Shadow Brokers, who used *Eternal Blue*, an exploit tool originally developed by the "US National Security Agency (NSA)" and was purportedly stolen by other actors. It exploited a vulnerability in Windows OSs and asked for a ransom ranging from "hundreds to ten thousands of dollars" in the form of Bitcoin (similar to *CTB-Locker* and *Petya*) (Francis, 2016; Mullin & Lake, 2017; *Symantec*, 2017).

There have been rampant leakages of user data from Facebook, an incident that saw data leak, including "profile names, Facebook ID numbers, email addresses, and phone numbers" of nearly 533 million users (Newman, 2021). Simultaneously, various government sectors in Saudi Arabia and the US were targeted by hackers from Iran. Further, incidents of Russian and Chinese hacking activities are increasingly evident as well. The indictment of Russian intelligence agents and hackers and "attempts to penetrate… military, government and defence industry networks" by Chinese authorities were all of greater concern recently.

Other major incidents include the release of hacking tools, originally of National Security Agency or NSA by the *Shadow Brokers*; the Defense Ministry email accounts being illegally accessed by Russian espionage groups; cyber espionage attempts by Chinese hackers on various sectors in the US, Europe and Japan; spear-phishing by North Korean-based Lazarus Group on US-based defence contractors; the hacking incident of Ireland-based EirGrid by rival state-sponsored hackers; Israeli-Iranian cyber attacks and counter-attacks were all (CSIS, 2021). The history and evolution of cyber threat thus transformed and enabled the transition to the securitization of cyberspace. Even then, the pertinent issue of vulnerabilities and potential threats from cyberspace has but had a manifold increase, often having greater implications in restructuring the World Order (Dilipraj, 2016, 139–140).

THEORETICAL DIMENSIONS: CYBERSPACE AND SECURITY IMPLICATIONS

The traditional theorization of geopolitics, political economy, comparative politics or foreign policy, all of them converge with the cybersecurity narrative. But, new and unique areas have developed as well. Rosecrance (1996, 1999) elucidates on aspects like urban concentrations, global citizenship and issues of global citizenship and anarchy. Choucri (2000) has argued that paradigmatic transitions and transformations are happening. The work enunciates on three major aspects: globalization, worldwide electronic connectivity and emergent practices in knowledge networking. These works laid the foundation of all major meta- and/or grand-narratives on cyber governance, security and international relations for the twenty-first century. Moreover, the increasingly interdisciplinary and cross-disciplinary scholarly works, cybersecurity in international relations has become the buzzword in the social sciences disciplines, converging with other disciplines like computer sciences, media studies, biological sciences, chemical sciences and other related disciplines as well.

Osiander (2001) elucidates on how the seventeenth-century narrative of "Westphalia" is actually an end-product of the nineteenth- and twentieth-century "fixation on the concept of sovereignty" (Osiander, 2001, 251). It ended with the signing of treaties to end both the Thirty Years' War and the Eighty Years' War. The Eighty Years' War (1568–1648) was a struggle between the "universalists and particularists" that ended with the victory for the latter (Ibid., 252), and is evident in the struggle for power in cyberspace. The idea of exclusive sovereignty came into being, initially in the West, and later on to Africa and Asia, as an after-effect of colonization. Further, it led to the institutionalization of nation-states (*Boundless World History*, 2020) and "the right of all states to full independence" or territorial and political sovereignty over a particular land, people and governance. This was considered to be the "culmination of the anti-hegemonic struggle against the Hapsburg aspirations for a supranational empire" (Osiander, 2001, 252). This has vital in understanding state perceptions, actions and activities and complexities of cyber sovereignty.

An important aspect of cyber sovereignty in the Westphalian era can be understood by conceptualizing theoretical underpinnings of sovereignty. According to Nagan and Hammer (2004), sovereignty include: real or ritualized head of state; absolute or unlimited control or power; political

legitimacy; political authority; self-determination or national independence, governance and constitutional order; a criterion of jurisprudential
validation of all law (within respective territory); juridical personality
of sovereign equality; the symbol of recognition; formal unit of legal
system; powers, immunities or privileges; jurisdictional competence to
make and/or apply law; constitutive processes or basic governance
competencies (141–187).

Henry Kissinger in his book *World Order* (2014), "each state was
assigned the attribute of sovereign power over its territory". Each would
acknowledge the domestic structures and religious vocations of its fellow
states as realities and refrain from challenging their existence (Kissinger,
2014, 9). Further, the principles "of the Westphalian model for a multi-
polar order of sovereign states" became an internationally accepted norm
(Ibid., 106).Interestingly, the European idea of sovereignty had certain
exceptions as in cases of colonizing parts of Africa and Asia against their
will.

The volume of attacks on nation-states, with huge implications like in
Estonia (in 2007), Georgia (in 2008), Kyrygystan (in 2009), Iran (in
2010), US elections (in 2016), INS Sindhuratna and defence institutions (in 2017) and Georgia (in 2019) have all changed the narrative
(Reghunadhan, 2018a, 37–50). Further, the threats to critical national
infrastructure (CNI), internet networks, electrical grid, government
websites, ministry portals, twitter accounts of various institutions, politicians, etc. The terms like cyberwarfare, cybercrime, cyberespionage and
cyberterrorism have become buzzwords in the study of cyberspace in
social sciences, international relations and other interdisciplinary studies
as well. Scholars like Brown and King (2000) elucidate on constructivist understanding, and problem-based learning utilized by academia;
and apply these concepts with International Relations, as in high-tech
problem-based simulation (Brown & King, 2000, 245–254).

Koehn and Rosenau (2002) elaborate on the framework that provides
relevance of transnational competence to the dynamics that transform;
civil society networks that permeate domestic–foreign frontiers. Additionally, it details into the interorganizational knowledge generation, aggregation, partnerships, interpersonal/intercultural interactions; transterritorial
networks and projects; worldwide skill revolution; and governance (105–
127). Asal (2005) focuses on the teachings of International Relations
theory, classical realism in cyberspace. The application of the "prison-
er's dilemma" to the nth degree, and diplomacy in something dealt with

in much detail (359–373). Gilboa (2005) details on the origins and development of Cable News Network (CNN) effect hypothesis; and the utilization of a new agenda for more useful approaches towards different effects of global communication (325–341).

Rozanova (2016) emphasizes on digitalization, technological competition and market structures, information technology platforms and monopoly or oligopoly, network market transformation in contemporary digital era, indivisible multimedia area, concentric (ring) market and/or the facets of a centre–periphery model (13–20). Olevsky (2016) deals with various aspects of cyberspace and global market, internationalization of business, manufacturing industry and technology, national competitiveness (17–26). Dmitriev and Korolev (2017) focused on cybersecurity, Intellectual Property Rights (IPR), technology transfer and international trading system. It provided the framework to inculcate emerging economic strategies, economic policies, economic protectionism and foreign policy (25–36).

Zvonova and Kuznetsov (2018) detailed on the aspects of digital currency, cryptocurrency, bitcoin, virtual market space, global financial cybernetic system devoid from control of state organizations and transnational corporations (5–16). Arbatov (2019) deals with prominent mainstream narratives of cyberwarfare and the impact on arms control, revisionist understanding of military technologies, security implications of arms control, notions of parity and strategic stability. This has been the prominent trend and framework of security studies area (5–16). The cyber attacks and related incidents at Estonia (2007), Georgia (2008), Kyrygystan (2009), Iran (2010), US elections (2016) and defence institutions and submarine, warship fleet of India (2012–19) have all changed the narrative (Carayannis et al., 2018; Huskaj & Wilson, 2020; Monnappa, 2017; Siddiqui, 2020; Singh, 2012; Valeriano et al., 2018).

Whyte (2020) enunciates on the emerging implications of *cyber sovereignty* and *security* in democracies as well as issues in elections. This was based on the implications of state-led intervention in elections of other countries. This inevitably led to the sustenance of Trumpism and the support to white-nationalist tendencies within the political establishment in the last four years. Whyte calls it the cyber-enabled information warfare (IW) campaigns that have led to the "hacking" of democracies in the West, drawing parallels to the "democratic functionality" within the disciplines like political science, international relations and communication studies (1–17). This had more implications on the agency and function of not just citizens, but has implications on international relations and global politics.

Staak and Wolf (2019) have emphasized on cybersecurity on electoral cycle, the issues of vulnerabilities, disinformation strategies, mitigation measures and the need for interagency collaboration. The case studies of developed and developing countries in Europe, with the emergence and institutionalization of digital developments have provided much greater detail (Staak & Wolf, 2019). Meanwhile, transdisciplinary theoretical frameworks like cyberfeminism, cyberfeminism 2.0 and post-cyberfeminism have evolved through works of scholars like Donna Haraway, Sadie Plant, Mia Consalvo and Caroline Elbaor (Reghunadhan, 2018b, 41–50).

International Political Economy of Cyber Threat Landscape

In the Era of Digitalisation and World Wide Web (*www*), the capital accumulation has become much more invisible and the related understanding of exploitation, as well as appropriation of labour by New (Corporate) Capitalists, has not reached the mainstream narratives and discourses. Fuchs and Sevignani (2013) argue about "digital labour" or "digital work" in relation to the cyberspace as "a dominant form of capital accumulation model" of the MNCs. They argue that "the exploitation of the unpaid labour" of netizens like content creation, blogging, use of social media, microblogging, the platforms of decentralized forms of knowledge dissemination like wikis are all freely exploited (charged) by the Internet and social media companies (237–293).

According to scholars like Marx, the major source of capital accumulation for the bourgeois class is the exploitation and appropriation of labour by the population. Historically, this has been visible and apparent to theorists in understanding the underlying mechanism of the oppressive nature of capitalist systems. Thus, the exploitation of just the working or labour class, has transformed virtually into everyone who has connected to the Internet or inadvertently the digital world. In the era of the Internet of Things (IoT), global (inter and intra) connectivity and cyber technologies, the other most important commodity is the data. This has shown how Big Data and data analytics, (and in the future) combined with Artificial Intelligence (AI) would control the world and the related decision-making process in the world. Besides, the commodification of data and targeted advertising enables the companies to take control over the netizens and in turn sell the surplus value of exploited unpaid labour for a higher price, often to the highest bidder (Marx, 1867).

The use-value of the "commodified data" and its exchange-value is determined by the market, that which is often monopolized or oligopolized by a handful of these MNCs. The data of the individual or the "population user" or netizen provides billions (if not trillions) of capital to the MNCs. The access to these data is free for these companies which sell it for the largest bidder. Thus, a commodification of these data and its market that is often inaccessible to its original owners has arisen. This has led to capital accumulation for the MNCs by exploitation of the netizen and their data. This has also resulted in the alienation of the user in the cyberspace. The "economic geographies" in these emerging datascapes have but transformed the spatializations and their materialization in the current society. An increased control and monopolization of information by corporates and MNCs have led to hierarchical re-ordering of the knowledge (or informational) economy that leads to (in)direct and (in)visible exploitation of the population, degrading of freewill as well as decision-making.

The exchange-value of "information, and the industries of production, storage and retrieval, circulation [as well as] consumption of data" have been determined by a market that has the "(in)visible hand" of the MNCs. Interestingly, Smithian as well as Ricardian considerations of market functioning and processes fails and so does the modern conceptions of neoliberalism and free society. The economic base and related superstructure entail a new form of capital accumulation that stems from exploitation and looting of personal and other related data, besides alienation of the user as well. This surplus value created through the unpaid digital labour/work and the tacit commodification of data and privacy of the netizens, by the new capitalist (corporate) minority, is running global operations on a scale that has been hitherto heard of (Turner, 2018, 53–62).

Interestingly, though the target of exploitation, unlike the (conventional sense of) working class has been transferred to the general population as a whole, and thus increases the "target value" for these large MNCs. This has entered into a consistent phase of newer forms of capital accumulation; what some argues as "surveillance capitalism". Besides, the methods and phases of exploitation have but transformed due to development of ICTs. Surveillance capitalism tends to have full control over the thought(s), process(es) as well as behaviour(s) of the "population users" or netizens, on a massive scale (Ibid.). In October 2018, according to the World Payment Report 2018, the circulation of money through the e-wallet provided by the top six global companies in 2017 was 41.8 billion USD out of the total 359.48 billion USD. This accounts for 8.6% of the total global non-cash transactions. The report enunciates that more than 29 billion USD, accounting for approximately 71% of these transactions were through Big Tech companies "such as Google, Amazon, Facebook, Apple, Alibaba, and Tencent" (Capgemini & BNP Paribas, 2018; Nian, 2017).

The global non-cash transactions mainly depend on digital transactions and authentication, that primarily ascribes to the use of Internet and mobile apps, as well as the "circulation" of data and its commodification that are controlled by these Big Tech MNCs. This has shown the emergence of new forms of transactions and transmission of exchange of value among peers, showing a 10.1% increase in total cash flow. In 2016–17 forecasts, the global growth was around 11%, with Emerging Asia with 29%, Central Europe, Middle East and Africa (CEMEA) with 17.4% and Mature Asia–Pacific with 10.1%. By 2021 the global non-cash transactions are expected to reach 876.4 billion USD, and the top three regions with highest global non-cash transactions would be at Emerging Asia (250.7 billion USD), North America (211.7 billion USD) and Europe (151.1 billion USD) (Ibid.).

Besides, the traditional markets and conventional financial systems of the modern-industrialized societies in the US and West European countries are able to maintain high global non-cash transactions. But the mature market regions are likely to fall behind in terms of growth rates in developing regions. The developing market regions that comprise of Emerging Asia and CEMEA are expected to be the engines of growth. But these markets would be monopolized by the few Big Tech MNCs. In China, this has been a huge debate, particularly due to the direct linkages to companies. Besides providing more control over the population, the internet revolution provided more economic wherewithal to the Chinese State, and to the population. The later years saw the emergence and development of four major internet-cum-tech companies Tencent (in 1998), JD.com (in 1998), Alibaba (in 1999) and Baidu (in 2000); all of whom as of by May 2018 accounted for a combined 19.49% of the total global market value and related shares (*Alibaba Group*, 2018; *Baidu*, 2018; *JD.COM*, 2018; Molla, 2018; *Tencent*, 2018).

Interestingly, it was reported that Alibaba has broken into the top ten global companies in terms of share value, immediately after it broke the record (at 25 billion USD) for the largest Initial Public Offering (IPO) by a company in the history of share market (Zucchi, 2019). Currently, China accounts for more than 40% of the total digital commercial transactions (e-commerce), amounting to approximately 3.6 trillion USD around the globe. This has been mostly controlled by Big Tech companies like Alibaba and Tencent, who control, store, circulate and sell the data and information of the population (Jiang & King-wa, 2018, 372–392). In India, the State narrative is that it has taken various initiatives to bridge digital divide and increase accessibility to the common population (*MeiTY*, 2019). But does this ordain much necessary concentration of power and capital towards MNCs like Facebook, Google and Amazon, as well as their activities in India is something that has not been part of the mainstream discussion, narratives or studies.

A research paper published by ESET, an antivirus and internet security provider based in Slovakia, reveals that a wide range of operating systems have been and are still vulnerable to attacks. These include Apple OS X, Microsoft Windows and Linux (Bilodeau et al., 2014). In the same year, Centre for Strategic & International Studies (CSIS) came out with a report, funded by US-based McAfee Inc. estimated that cyber threats cost 445 billion USD annually, amounting to around one per cent of the global income. Stewart Baker, who co-authored the report and is a former policy official at the Department of Homeland Security opined that cyber-related attacks "are big, and they're growing" (CSIS, 2021; Nakashima & Andrea, 2014).

In 2014, the cost of damages inflicted by cyber attacks was estimated to be around 400 billion USD, with attacks prominently originating not just from Russia and China, but Nigeria, Brazil, Vietnam (Rayman, 2014) and various countries in the Middle East as well. According to a study by IBM Security, an estimated 1.5 million cyber attacks take place annually; amounting to 4000 attacks per day. It estimates an average of 16,856 cyber attacks on businesses, amounting to 46 cyber attacks on businesses per day with a success rate of half a per cent. That amounts to at least four dozen successful cyber attacks per year on businesses per se (*CSI: Cyber*, 2015). The top ten countries that were affected include Taiwan (24%), the US (16%), Indonesia (nine per cent), Mexico (eight per cent), Malaysia (six per cent), Israel (five per cent), Italy (five per cent), Vietnam (two per cent), France (two per cent) and Spain (two per cent) and had major economic as well as national security dilemma for these nations (Daniel, 2016).

According to the report by Accenture titled *Cost of Cyber Crime Study*, it is estimated that there is a 27.4% increase in cybercrimes every year, with average number of successful breaches increasing three times per year, with ransomware attacks alone doubling in frequency; from 13 to 27%. The incidents of *WannaCry* and *Petya* affected thousands, if not millions of devices and disrupted public services like health, public distribution system, Supervisory Control and Data Acquisition (SCADA) systems and the like. The customer records of Equifax (consumer credit reporting agency) amounting to more than 100 million was stolen, making information theft one of the most expensive crimes of all time (Accenture, 2016; Reghunadhan, 2020). Each successful cyber attack could amount to quite a large amount of money, more importantly, loss of data and research documents of huge economic and strategic value. It adopts non-physical aspects of cyberspace and thus is considered to be in a paradigm which is beyond border, time and is considered to be anonymous; in a relative sense. In 2018, the emergence of cyber threats pertinent to issues of stealing cryptocurrencies has an estimated cost of 1.1 billion USD (Rick, 2018).

FBI reported that cybercrime created a combined total loss of over 780 million USD, averaging 3,000 USD per complaint. These include categories like romance scammers amounting to (12,000 USD per complaint), auto scammers amounting to (3,600 USD per complaint), real estate rental scams amounting to (1,800 USD per complaint) and FBI scammers amounting to (700 USD per complaint) (*CSI: Cyber*, 2015). The FBI also reports that there is a very low probability of 1 in 20,000 to capture the real perpetrator in case of cyber-related attacks. This is even more cause for problem, particularly in case of more knowledge, know-how, accessibility to systems, outdated or obsolete passwords, mechanisms and procedures. Accordingly, it estimates that 60% of thefts (personal or identity) occur due to this (FBI, 2018). The expansion of the Internet has but enabled an increased access to different locations, objects or even individuals around the world, while enhancing skill-sets and know-how for various cyber criminals, groupings and organizations. This has increased the complexities and security concerns, particularly for preparedness in protecting against increased threats and attacks even at the individual level. A large number of cyber attacks are undertaken by private individuals, groups and/ or organizations particularly directed against private individual users as well as businesses. It takes place in varying forms in the likes of identity theft, piracy, hacking, cyber-bullying, transaction fraud, advance fee fraud, distribution of child exploitation or sexual abuse material, prohibited offensive or illegal content and the like.

In February 2016, a reported 81 million USD bank heist was uncovered, which transferred millions within hours. It subverted a high-security closed computer network that was only accessible between member's banks in the world (Zetter, 2016). Luckily for the authorities the target of transferring one billion USD was prevented. There have been increasing incidences of attacks targeting women and children (Rabbi, 2017). The major problem with categorization of cyber attacks as cyberwarfare, cyber terrorism, cyber espionage or cybercrime is the identification of real actor(s) or perpetrator(s) and their motive(s) behind the attacks or series of attacks. Many a time a cyber attack would be based on a digital footprint left behind by the perpetrator, which could be planted, manipulated or even deceiving. Since the cyber attacks take place in a matter of seconds, the conventionally accepted form of investigation will not be able to identify and/or investigate the matter in a timely manner.

The Internet Security Threat Report (ISTR) by US-based security provider Symantec lists major causes of cyber attacks that include theft of data, improper use of data, phishing, spoofing or social engineering, accidental data loss, loss or theft of device, IT errors leading to data loss, network disruption or DDoS, extortion, blackmail or disruption as well as identity theft. These are mainly focused on economic aspects of the users mostly motivating these forms of attacks leading to multiple impacts on victims. In 2017, report by Symantec on global cybercrime cost estimated around billions, if not trillions of dollars with millions of victims. It is showing an increase each year and causing greater damage to individuals, organizations as well as nations. It opined that "cyber criminals caused unprecedented levels of disruption with relatively simple IT tools and cloud services". The innovation, technological revolution as well as increased level and rate of sophistication have made cyber attacks a greater challenge as well as threat (*Symantec*, 2017). Symantec actually warns against the possibility of AI-enabled cyber attacks that can possibly create "an explosion of network penetration, personal [information] theft and an [advanced] level... viruses" in the near future if not now (Mandavia, 2017; *Techopedia*, 2018a).

In nations like the US, Australia, Germany, France, UK and in powerful "developing nations" like China and India, cyber attacks are specifically dealt with, particularly in relation to investigation and prosecution. While the other developing nations, under developed nations and/or least developed Countries (LDCs) like African nations, most of the Asian countries and South American nations have lesser expertise. It thus requires requisite training, tools and skill to deal with the emerging threats from cyber attacks, both in terms of personnel and technological know-how to deal with cyber attacks, besides the lack of legislative wherewithal in these nations. The increasing assertiveness of China, Russia, Iran, North Korea and others in the cyberspace and its increased security posture has actually exacerbated the already increasing cyber attacks in the cyberspace, with citizens tending to be the weakest link exposed to the cyber attacks originating from these nations.

CONCLUSION

The globalization process that the networks and cyberspace have created and initiated have created a borderless, barrier-less platform and/or domain for the cyber threat landscape. The emergence of automation, multiple time zones, ease of accessibility to computer systems have all increased the incidence of cyber-related activities and with-it glaring dangers. The timeless and borderless nature of the cyber environment, where only digital information and data is disseminated and exchanged, makes it difficult to know who or what may be operating a device or computer. The more skilled the user, the relatively easier it is to cover the digital footprint (tracks) in cyberspace and more difficult to investigate as well as to get evidence.

Further, cybercriminals are also taking advantage of the internationalization of cyberspace by networking with cybercriminals as well as creating mafias and gangs. The technical know-how and sophistication of a cybercriminal and the lack of updated viable legislative-cum-investigative mechanisms, legislation and practices enhance the possibilities and often motivate the person to commit (digital and resultant physical) violence more often at a larger magnitude. This, combined with networking and grouping different gangs and groupings with sophistication, enlarges the threats and possible concentrated attacks over institutional and/or non-institutional actors in cyberspace. Electronic devices, including computers, and the integration provided from telecommunications networks, have increasingly led to interconnectivity, interoperability, scalability and coordination between various systems across the world. It has enhanced the instances of hacking and an exponential increase in adversarial activities affecting host nations and individuals.

The international aspects of cyber threats have further created complex but variegated difficulties for various nations and stakeholders, which is much more difficult to mitigate and prevent, at least in terms of the aftereffects they have led. Historically, cyber threat has come in many forms and is most often associated with its most advertised forms *vis-à-vis.*, spamming, malicious software(s) like viruses, worms, trojans and/or malware, from quarters, often expected. The rise of DoS, DDoS, ransomware, malware and trojan programs that target and attack governments, businesses, organizations are becoming more prevalent, sophisticated and often cost billions (if not trillions) of dollars of damages in relation to loss as well as recovery.

Since physical borders don't bind cyberspace, the accessibility and perpetuation of tools by cyber hackers from anywhere is always a possibility and a potential vulnerability. This has, in turn, destabilized and unbalanced the conventional understanding of power politics and its perpetuation, mainly in the aftermath of globalization led and perpetuated by digitization. Unlike the physical-cum-territorial measures undertaken to offset potential and possible threats, the transboundary nature of cyberspace brings in multi-dimensional threat perceptions that are much complex to be dealt with. It has had created a greater problem in the socio-political and economic conditions of individuals in particular, and society in general. The need to prevent any exacerbation of issues, particularly through the misuse of information and communication technologies (ICTs) for malignant and/or anti-social activities, has to be dealt with greater coordination between different actors, State and non-State but with larger collaboration among netizens as well.

There arise challenges for State actors, both at the individual and collective level, to deal with the impact of cyber attacks. Unlike the traditional threat perceptions, related securitizations, the existing security architecture often becomes inept at dealing with the increasing threats from cyber attacks to critical national infrastructure, particularly the inability of legal systems as well as the response mechanisms. States need to formulate strategies, particularly for the development of frameworks capable of enabling support for policies, activities and implementation of measures to deal with the impact of cyber attacks as a prerequisite. The legal and institutional support, particularly in regard to capacity building and increased coordination and information sharing of multiple agencies, actually requires greater collaboration and cooperation between different actors. In this regard, the role of government remains important, as it enables the development of proactive but relatively effective legislative mechanisms, procedures and processes to deal with the threat of cyber attacks. This has to be supplemented by ensuing information sharing and cooperation between institutional actors.

References

Accenture. (2016). *Editing the uneditable blockchain: Why distributed ledger technology must adapt to an imperfect world.* Retrieved December 17, 2019, from https://www.accenture.com/cn-en/insight-editing-uneditable-blockchain

Alibaba Group. (2018). *Company overview.* Retrieved March 27, 2019, from https://www.alibabagroup.com/en/about/overview

Arbatov, A. G. (2019). Dreams and realities of arms control. *World Economy and International Relations, 63*(11), 5–16.

Armed Forces Communications and Electronics Association (AFCEA). (2018). *The Impact of Federal German Intelligence Service Bundesnachrichtendienst (BND) Project RAHAB and Chaos Computing Congresses (CCC) impact on the Future of Computer-Network Mediated Espionage: Cuckoo's Egg Prequel or Perfect Storm?* Retrieved August 21, 2019, from https://www.afcea.org/committees/cyber/documents/impactofbndprojectrahabandcccconthefutureof computer-networkmediatedespionage-cuckooseggpreque.pdf

Asal, V. (2005). Playing games with international relations. *International Studies Perspectives, 6*(3), 359–373.

Baidu. (2018). *Company overview.* Retrieved July 7, 2020, from http://ir.baidu.com/phoenix.zhtml?c=188488&p=irol-homeprofile

Beal, V. (2018). *Zeus.* Retrieved July 7, 2020, from https://www.webopedia.com/TERM/Z/zeus.html

Benson, R. (2014). *Trust and partnership: Strategic IT management for turbulent times.* John Wiley & Sons.

Bilodeau, Bureau, O. P.-M., Dorais-Joncas, J. C. A., Léveillé, M.-E. M., & Vanheuverzwijn, B. (2014). *Operation Windigo, ESET.* Retrieved May 11, 2018, from https://www.welivesecurity.com/wp-content/uploads/2014/03/operation_windigo.pdf

Boundless World History. (2020). *Nation-States and Sovereignty.* Lumen. Retrieved December 11, 2020, from https://courses.lumenlearning.com/boundless-worldhistory/chapter/nation-states-and-sovereignty/

Bradley, T. (2016). *Sub7Trojan/Backdoor.* Retrieved May 7, 2019, from https://www.lifewire.com/sub7-trojan-backdoor-2486800

British Broadcasting Corporation (BBC). (2021). *Shamoon virus targets energy sector infrastructure*. Retrieved February 7, 2021, from https://www.bbc.com/news/technology-19293797

Britz, M. (2012). *Computer forensics and cyber crime: An introduction* (2nd ed.). Dorling Kindersley India.

Brookes, T. (2011). *6 computer viruses that changed the world*. Retrieved July 13, 2018, from https://www.makeuseof.com/tag/6-computer-viruses-changed-world/

Brown, S. W., & King, F. B. (2000). Constructivist pedagogy and how we learn: Educational psychology meets international studies. *International Studies Perspectives, 1*(3), 245–254.

Capgemini and BNP Paribas. (2018). *World payments report 2020*. Retrieved February 2, 2020, from https://worldpaymentsreport.com/resources/world-payments-report-2020/

Carayannis, E. G., Campbell, D. F. J., & Efthymiopoulos, M. P. (Eds.). (2018). *Handbook of cyber-development, cyber-democracy, and cyber-defense* (pp. 1–1089). Springer International Publishing.

Carnegie Mellon University (CMU). (1999). Happy99.exe trojan horse. *CERT incident note CA-1999–02*. Retrieved June 17, 2018, from http://www.cert.org/incident_notes/IN-99-02.html

Centre for Strategic and International Studies (CSIS). (2021). *Significant cyber incidents since 2006*. Retrieved August 11, 2021, from https://www.csis.org/programs/strategic-technologies-program/significant-cyber-incidents

Cerulus, L. (2019). How Ukraine became a test bed for cyberweaponry. *Politico*. Retrieved December 30, 2020, from https://www.politico.eu/article/ukraine-cyber-war-frontline-russia-malware-attacks/

Chen, T. M. (2003). Trends in viruses and worms. *The Internet Protocol Journal, 6*(3).

Choucri, N. (2000). Introduction: CyberPolitics in international relations. *International Political Science Review, 21*(3), 243–263.

Cluley, G. (2009). GhostNet: Who is really behind it? *NakedSecurity*. Retrieved May 7, 2018, from https://nakedsecurity.sophos.com/2009/03/29/ghostnet/

Cluley, G. (2010). *A short history of Christmas malware*. Retrieved May 13, 2018, from https://nakedsecurity.sophos.com/2010/12/15/christmas-malware-short-history/

Cobb, S. (2014). Botnet malware: What it is and how to fight it. Retrieved May 13, 2018, from https://www.welivesecurity.com/2014/10/22/botnet-malware-fight/

Computer Hope. (2017). *Cridex malware*. Retrieved May 7, 2018, from https://www.computerhope.com/jargon/c/cridex-malware.htm

Cottam, H. (2020). *Welfare 5.0: Why we need a social revolution and how to make it happen*. UCL Institute for Innovation and Public Purpose (IIPP).

CSI: Cyber. (2015). *These cybercrime statistics will make you think twice about your password: Where's the CSI cyber team when you need them?* Retrieved June 1, 2018, from http://www.cbs.com/shows/csi-cyber/news/1003888/these-cybercrime-statistics-will-make-you-think-twice-about-your-password-where-s-the-csi-cyber-team-when-you-need-them-/

Dalakov, G. (2018). *First computer virus of Bob Thomas*. Retrieved February 17, 2019, from http://history-computer.com/Internet/Maturing/Thomas.html

Daniel, C. I. D. (2016). *Large CCTV botnet leveraged in DDoS attacks*. Retrieved February 10, 2018, from https://blog.sucuri.net/2016/06/large-cctv-botnet-leveraged-ddos-attacks.html

Dear, B. (2017). *The friendly orange glow: The untold story of the PLATO system and the dawn of cyberculture* (1st ed.). Pantheon Books.

Dilipraj, E. (2016). Challenges of cybersecurity to nuclear infrastructure. *Defence and Diplomacy Journal*, 5(3), (April-June), 139–140.

Dittrich, D. (1999). *The "stacheldraht" distributed denial of service attack tool*. Retrieved January 17, 2018, from https://staff.washington.edu/dittrich/misc/stacheldraht.analysis

Dmitriev, S. S., & Korolev, I. S. (2017). Contours of emerging U.S. foreign trade program. *World Economy and International Relations*, 61(10), 25–36.

Editor. (2016). Flashback friday: SQL slammer. *WeLiveSecurity*. Retrieved December 17, 2018, from https://www.welivesecurity.com/2016/09/30/flashback-friday-sql-slammer/

Eyerys. (1990). *1260, The first polymorphic virus*. Retrieved June 7, 2018, from https://www.eyerys.com/articles/timeline/1260-first-polymorphic-virus#event-a-href-articles-timeline-facebook-and-billion-userfacebook-and-a-billion-user-a

Federal Bureau of Investigation (FBI). (2018). *Cyber crime*. Retrieved May 15, 2019, from https://www.fbi.gov/investigate/cyber

Francis, R. (2016). The history of ransomware. Retrieved June 12, 2018, from https://www.csoonline.com/article/3095956/data-breach/the-history-of-ransomware.html#slide1

Fruhlinger, J. (2017a). *Petya ransomware and NotPetya malware: What you need to know now*. CSOIndia. Retrieved May 17, 2018, from https://www.csoonline.com/article/3233210/petya-ransomware-and-notpetya-malware-what-you-need-to-know-now.html

Fruhlinger, J. (2017b). *What is ransomware? How it works and how to remove it*. Retrieved May 17, 2018, from https://www.csoonline.com/article/3236183/ransomware/what-is-ransomware-how-it-works-and-how-to-remove-it.html

Fruhlinger, J. (2017c). *What is stuxnet, who created it and how does it work?* CSOIndia. Retrieved May 17, 2021, from https://www.csoonline.com/art icle/3218104/what-is-stuxnet-who-created-it-and-how-does-it-work.html

Fuchs, C., & Sevignani, S. (2013). *What is digital labour? What is digital work? What's their difference? And why do these questions matter for understanding social media?* Retrieved May 17, 2018, from http://citeseerx.ist.psu.edu/vie wdoc/download?doi=10.1.1.913.5614&rep=rep1&type=pdf, 237–293

Garnaeva, M., Wiel, J. V. D., Makrushin, D., Ivanov, A., & Namestnikov, Y. (2015). *Overall statistics for 2015.* SecureList. Retrieved February 27, 2018, from https://securelist.com/kaspersky-security-bulletin-2015-overall-statistics-for-2015/73038/

Garretson, C. (2007). *Storm: The largest botnet in the world?* NetworkWorld. Retrieved March 21, 2018, from https://www.networkworld.com/article/ 2286172/storm--the-largest-botnet-in-the-world-.html

Gilboa, E. (2005). Global television news and foreign policy: Debating the CNN effect. *International Studies Perspectives, 6*(3), 325–341.

González, R. J. (2017). Hacking the citizenry?: Personality profiling, 'big data' and the election of Donald Trump. *Anthropology Today, 33*(3), 9–12.

Gross, M. J. (2011). Cyber-espionage campaign and intellectual-property bonanza. *Vanity Fair.* Retrieved August 23, 2018, from https://www.van ityfair.com/news/2011/09/operation-shady-rat-201109

Harmon, A. (1998). *Hacking theft of $10 million from Citibank revealed.* Retrieved July 12, 2019, from http://articles.latimes.com/1995-08-19/bus iness/fi-36656_1_citibank-system

Hayashi, K. (2010). *Trojan.Archiveus.* Retrieved July 13, 2019, from https:// www.symantec.com/security_response/writeup.jsp?docid=2006-050601-094 0-99

Hipponen, M. (2011). *Brain virus.* Retrieved February 7, 2019, from https:// campaigns.f-secure.com/brain/virus.html

Howard, R. (Ed.). (2010). *Cyber fraud: Tactics, techniques and procedures.* Auerbach.

Huskaj, G., & Wilson, R. L. (2020). An anticipatory ethical analysis of offensive cyberspace operations. In *Proceedings of the 15th International Conference on Cyber Warfare and Security,* ICCWS 20202020, 512–520

INTERPOL. (2020a). *Cyber crime.* Retrieved January 11, 2021, from https:// www.interpol.int/Crime-areas/Cybercrime/Cybercrime

INTERPOL. (2020b). *Online safety.* Retrieved January 11, 2021, from https:// www.interpol.int/Crime-areas/Cybercrime/Online-safety/Ransomware

Isaak, J., & Hanna, M. J. (2018). User data privacy: Facebook, Cambridge Analytica, and privacy protection. *Computer, 51*(8), 8436400, 56–59.

JD.COM. (2018). *Our history.* Retrieved May 11, 2019, from http://corporate. jd.com/ourHistory

Jiang, M., & King-wa, F. (2018). Big data, big brother, big profit?. *Policy & Internet, 10*(4), 372–392.

Johnson, A. L. (2011). *W32.Duqu: The precursor to the next stuxnet*. Broadcom. Retrieved June 17, 2019, from https://community.broadcom.com/symant ecenterprise/communities/community-home/librarydocuments/viewdocum ent?DocumentKey=933c68f1-6ee7-473e-9eb6-6c8459f790f2&Community Key=1ecf5f55-9545-44d6-b0f4-4e4a7f5f5e68&tab=librarydocuments

Jones, S. (2015). *Plato, encyclopedia britannia*. Retrieved May 5, 2018, from https://www.britannica.com/topic/PLATO-education-system

Kaimba, B. (Ed.). (2017). *Tanzania cyber security report 2017: Demystifying Africa's cyber security poverty line*. Serianu. Retrieved May 11, 2018, from https://kabolik.com/downloads/Tanzania%20Cyber%20Security%20R eport%202017.pdf

Kaspersky. (2012). *Gauss: Nation-state cyber-surveillance meets banking Trojan*. SecureList. Retrieved May 23, 2019, from https://securelist.com/gauss-nat ion-state-cyber-surveillance-meets-banking-trojan-54/33854/

Kaspersky. (2018). *Zeus virus*. Retrieved February 9, 2020, from https://usa.kas persky.com/resource-center/threats/zeus-virus

Kelty, C. (2011). *The morris worm*. Retrieved June 1, 2018, from https://limn. it/the-morris-worm/#edn1

Kissinger, H. (2014). *World order*. Penguin Random House.

Knowbe4. (2018). *Ransomware knowledgebase. Koler.a*. Retrieved May 10, 2019, from https://www.knowbe4.com/kolera

Koehn, P. H., & Rosenau, J. N. (2002). Transnational competence in an emergence epoch. *International Studies Perspectives, 3*(2), 105–127.

Krebbs, B. (2016). *KrebsOnSecurity hit with record DDoS*. Retrieved May 3, 2019, from https://krebsonsecurity.com/2016/09/krebsonsecurity-hit-with-record-ddos/

Kushner, D. (2013). *The real story of stuxnet*. IEEE. Retrieved July 10, 2018, from https://spectrum.ieee.org/telecom/security/the-real-story-of-stuxnet

Lyons, M. J. (2000). Infamous hacker tracked to Latvia. *Baltic Times*. Retrieved October 9, 2019, from https://www.baltictimes.com/news/articles/640/

Mandavia, M. (2017). *Security software specialist symantec warns of AI-enabled cyber crimes*. Retrieved August 7, 2018, from https://economictimes.ind iatimes.com/tech/internet/security-software-specialist-symantec-warns-of-ai-enabled-cyber-crimes/articleshow/61649215.cms

Marsan, C.D. (2008). *Where is Robert Morris now?* Retrieved May 13, 2019, from https://www.networkworld.com/article/2268914/lan-wan/where-is-robert-morris-now-.html

Marx, K. (1867). *Capital: A critique of political economy: Volume one*. Penguin.

Mathis, J. (2011). *Intego: Malware masquerades as flash installer*. Macworld. Retrieved May 21, 2018, from https://www.macworld.com/article/116 2496/intego_malware_masquerades_as_flash_installer.html

McAfee. (2021). *What is stuxnet?* Retrieved February 7, 2021, from https://www.mcafee.com/enterprise/en-in/security-awareness/ransomware/what-is-stuxnet.html

Meredith, S. (2018). *Here's everything you need to know about the Cambridge Analytica scandal*. Retrieved July 11, 2020, from https://www.cnbc.com/2018/03/21/facebook-cambridge-analytica-scandal-everything-you-need-to-know.html

Meskauskas, T. (2021). *How to avoid installation of NetWire and other remote access tools*. PCRisk. Retrieved March 17, 2021, from https://www.pcrisk.com/removal-guides/15614-netwire-rat

Ministry of Electronics and Information Technology (MeiTY). (2019). *National e-Governance Plan*. Retrieved July 7, 2020, from http://meity.gov.in/divisions/national-e-governance-plan

Molla, R. (2018). *Mary Meeker: China now has nine of the world's biggest internet companies—Almost as many as the U.S. Recode*. Retrieved August 11, 2020, from https://www.recode.net/2018/5/30/17385376/mary-meeker-kleinerslides-code-conference-chinese-tech

Monnappa, K. A. (2017). *Cyber attack targeting Indian Navy's submarine and warship manufacturer*. CysInfo. Retrieved June 17, 2019, from https://cysinfo.com/cyber-attack-targeting-indian-navys-submarine-war ship-manufacturer/

Moore, R. (2015). *Cybercrime: Investigating high-technology computer crime*. Routledge.

Mullin, G., & Lake, E. (2017). *Malicious virus: What is wannacry ransomware? Malware used to cripple NHS in 2017 cyber attack*. Retrieved June 17, 2020, from https://www.thesun.co.uk/tech/3562470/wannacry-ransomware-nhs-cyber-attack-hackers-virus/

Nagan, W. P., & Hammer, C. (2004). The changing character of sovereignty in international law and international relations. *Columbia Journal of Transnational Law, 43*, 141–187.

Nakashima, E., & Andrea P. (2014). *Report: Cybercrime and espionage costs $445 billion annually*. Retrieved May 7, 2019, from https://www.washingtonpost.com/world/national-security/report-cybercrime-and-espionage-costs-445-bil lion-annually/2014/06/08/8995291c-ecce-11e3-9f5c-9075d5508f0a_story.html?utm_term=.7e0f4278563f

Naraine, R. (2007). *Storm Worm botnet could be world's most powerful supercomputer*. Retrieved March 1, 2018, from http://www.zdnet.com/article/storm-worm-botnet-could-be-worlds-most-powerful-supercomputer/

52   R. REGHUNADHAN

Newman, L. H. (2021). What really caused Facebook's 500M-user data leak? *WIRED*. Retrieved March 27, 2021, from https://www.wired.com/story/facebook-data-leak-500-million-users-phone-numbers/#intcid=_wired-right-rail_6eb5d94f-c910-4324-a2a0-6ecd9b7a5754_popular4-1-reranked-by-vidi

Nian, Y. (2017). China's online retail sales account for 40 percent of global market. *China Global Television Network group (CGTN)*. Retrieved September 17, 2019, from https://news.cgtn.com/news/3d497a4e3341444e/share_p.html

Official Site of the State of New Jersey (NJCCIC). (2016). *NetWire RAT: NJCCIC threat profile*. Retrieved October 7, 2019, from https://www.cyber.nj.gov/threat-center/threat-profiles/trojan-variants/netwire-rat

Olevsky, G. M. (2016). Internationalization of business and national competitiveness. *World Economy and International Relations, 60*(12), 17–26.

Osiander, A. (2001). Sovereignty, international relations, and the westphalian myth. *International Organization, 55*(2), 251–287.

OVH. (2016). *The DDoS that didn't break the camel's VAC**. Retrieved July 11, 2019, from https://www.ovh.com/us/news/articles/a2367.the-ddos-that-didnt-break-the-camels-vac

PLATO History Foundation. (2010). *Perhaps the first denial-of-service attack?* Retrieved February 19, 2019, from http://www.platohistory.org/blog/2010/02/perhaps-the-first-denial-of-service-attack.html

Rabbi, A. R. (2017). *Women biggest victims of rising cyber crimes*. Retrieved March 21, 2018, from http://www.dhakatribune.com/bangladesh/crime/2017/09/21/women-biggest-victims-rising-cyber-crimes/

Radware. (2016). *DDoS attacks on DNS services*. Retrieved March 16, 2019, from https://security.radware.com/ddos-threats-attacks/threat-advisories-att ack-reports/dns-services-under-attack/

Radware. (2017). *History of DDoS attacks*. Retrieved March 15, 2019, from https://security.radware.com/ddos-knowledge-center/ddos-chronicles/ddos-attacks-history/

Rayman, N. (2014). The world's top 5 cybercrime hotspots. *Time*. Retrieved March 15, 2019, from http://time.com/3087768/the-worlds-5-cybercrime-hotspots/

Reghunadhan, R. (2018). Cyber threat landscape of digital India: A critical perspective. *Journal of Polity and Society, 10*(1&2), 37–50.

Reghunadhan, R. (2018). Gender inequality in cyberspace: A study on socio-technological implications to women in Kerala. *Women's Link Journal, 24*(4), 41–50.

Reghunadhan, R. (2020). Ethical considerations and issues of blockchain technology-based systems in war zones: A case study approach. In Krishnan, S., Balas, V. E., Julie, E. G., Robinson, Y. H., Balaji, S., & Kumar, R. (Eds.),

Handbook of research on blockchain technology (pp. 1–34). Academic Press (Elsevier).

Rick, D. (2018). *Crypto cybercrime thefts total over $1.1 billion in first six months of 2018.* Retrieved March 17, 2019, from https://www.newsbtc.com/2018/06/07/crypto-cybercrime-thefts-total-over-1-1-billion-in-first-six-months-of-2018/

Rosecrance, R. (1996). The rise of the virtual state. *Foreign Affairs, 75*(4), 45–61.

Rosecrance, R. (1999). *The rise of the virtual state.* Basic Books.

Rouse, M. (2011). *TDL-4 (TDSS or Alureon).* Retrieved December 17, 2019, from http://searchsecurity.techtarget.com/definition/TDL-4-TDSS-or-Alureon

Rozanova, N. M. (2016). Network competition as determinant of contemporary markets' configuration. *World Economy and International Relations, 60*(4), 13–20.

Schiller, C. A., Binkley, J., Harley, D. Evron, G., Bradley, T., Willems, C., & Cross, M. (2007). *Chapter 1—Botnets: A call to action.* Botnets, 1–27.

Shactman, N. (2008). Russian coder: I hacked Georgia's sites in cyberwar. *WIRED.* Retrieved May 5, 2019, from https://www.wired.com/2008/10/government-and/

Shehod, A. (2016). Ukraine power grid cyberattack and US susceptibility: Cybersecurity implications of smart grid advancements in the US. *Cybersecurity Interdisciplinary Systems Laboratory (CISL) Working Paper CISL# 2016–22.* Retrieved February 9, 2020, from https://web.mit.edu/smadnick/www/wp/2016-22.pdf

Shoch, J., & Hupp, J. (1982). The "worm" programs early experience with a distributed computation. *Communications of ACM, 25,* 172–180.

Siddiqui, H. (2020). Indian Navy's challenges: Countering Chinese naval activities in the Indian Ocean region. *Financial Express.* Retrieved January 11, 2021, from https://www.financialexpress.com/defence/indian-navys-challenges-countering-chinese-naval-activities-in-the-indian-ocean-region/1983186/

Singh, S. K. (2012). *Cyber crooks attack Indian armed forces.* Outlook. Retrieved February 5, 2019, from https://www.outlookindia.com/newsscroll/cyber-crooks-attack-indian-armed-forces/1679965

Spafford, E. (1989). The Internet worm program: An analysis. *ACM Computer Communications Review, 19* (January), 17–57.

Sponchioni, R., & Zhicheng Z. (2017). *Android.Simplocker.* Retrieved September 11, 2020, from https://www.symantec.com/security_response/writeup.jsp?docid=2014-060610-5533-99

Staak, S. V. D., & Wolf, P. (2019). *Cybersecurity in elections: Models of interagency collaboration.* International Institute for Democracy and Electoral Assistance

Stewart, J. (2007). *Storm worm DDoS attacks*. Retrieved March 12, 2019, from https://www.secureworks.com/research/storm-worm

Stroud, F. (2018a). *Cridex malware*. Retrieved March 14, 2019, from https://www.webopedia.com/TERM/C/cridex-malware.html

Stroud, F. (2018b). *Dridex malware*. Retrieved March 16, 2019, from https://www.webopedia.com/TERM/D/dridex-malware.html

Sullivan, J. (2000). *2 Arrested in bloomberg extortion case*. Bloomberg. Retrieved March 2, 2019, from http://www.nytimes.com/2000/08/15/business/2-arrested-in-bloomberg-extortion-case.html

Symantec. (2017). *Internet Security Threat Report (ISTR), vol. 22*. Retrieved May 12, 2021, from https://www.symantec.com/security-center/threat-report

Szor, P. (2005). *The art of computer virus research and defense*. Pearson Education Inc.

Techopedia. (2018a). *Spoofing*. Retrieved May 12, 2020, from https://www.techopedia.com/definition/5398/spoofing

Techopedia. (2018b). *Elk cloner*. Retrieved May 15, 2020, from https://www.techopedia.com/definition/25199/elk-cloner

Techopedia. (2018c). *Jerusalem virus*. Retrieved May 12, 2020, from https://www.techopedia.com/definition/27875/jerusalem-virus

Techopedia. (2020). *SQL slammer*. Janalta Interactive. Retrieved August 7, 2021, from https://www.techopedia.com/definition/27496/sql-slammer

TechTarget. (2009). *Conflicker*. Retrieved February 5, 2018, from http://whatis.techtarget.com/definition/Conficker

Tencent. (2018). *About*. Retrieved May 11, 2018, from https://www.tencent.com/en-us/

TrendMicro. (2017). *The michelangelo virus, 25 years later*. Retrieved October 9, 2018, from https://www.trendmicro.com/vinfo/us/security/news/cybercrime-and-digital-threats/the-michelangelo-virus-25-years-later

Turner, F. (2018). The arts at Facebook: An aesthetic infrastructure for surveillance capitalism. *Poetics, 67*, 53–62.

Ullah, F., Edwards, M., Ramdhany, R., Chitchyan, R., Babar, M. A., & Rashid, A. (2018). Data exfiltration. *Journal of Network and Computer Applications, 101*(C), 18–54.

Valeriano, B., Jensen, B., & Maness, R. C. (2018). *Cyber strategy: The evolving character of power and coercion* (pp. 1–306). Oxford University Press.

van Wyk, K. R. (1989). The lehigh virus. *Computers & Security, 8*(2), 107–110.

VanDerWerff, E., & Lee, T. B. (2015). *The 2014 sony hacks, explained*. Vox. Retrieved May 12, 2018, from https://www.vox.com/2015/1/20/18089084/sony-hack-north-korea

Waddel, K. (2016). *The computer virus that haunted early AIDS researchers*. Retrieved June 7, 2019, from https://www.theatlantic.com/technology/arc

hive/2016/05/the-computer-virus-that-haunted-early-aids-researchers/481
965/

Weyhrich, S. (2010). *Apple II history.* Retrieved June 17, 2018, from https://
apple2history.org/history/ah23/#08

Whyte, C. (2020). Cyber conflict or democracy "hacked"? How cyber operations
enhance information warfare. *Journal of Cybersecurity, 6*(1), 1–17.

WIRED. (1998). *Russian bank hacker sentenced.* Retrieved October 2, 2018,
from https://www.wired.com/1998/02/russian-bank-hacker-sentenced/

Zetter, K. (2010). Report strengthens suspicions that stuxnet sabotaged Iran's
nuclear plant. *WIRED.* Retrieved May 7, 2018, from https://www.wired.
com/2010/12/isis-report-on-stuxnet/

Zetter, K. (2011). Son of stuxnet found in the wild on systems in Europe.
WIRED. Retrieved May 7, 2019, from https://www.wired.com/2011/10/
son-of-stuxnet-in-the-wild/

Zetter, K. (2012). Meet 'flame,' The massive spy malware infiltrating Iranian
computers. *WIRED.* Retrieved July 7, 2019, from https://www.wired.com/
2012/05/flame/

Zetter, K. (2016). That insane, $81M Bangladesh bank heist? Here's what
we know. *WIRED.* Retrieved July 9, 2019, from https://www.wired.com/
2016/05/insane-81m-bangladesh-bank-heist-heres-know/

Zucchi, K. (2019). *Top 10 largest global IPOs of all time.* Investopedia. Retrieved
July 21, 2020, from https://www.investopedia.com/articles/investing/011
215/top-10-largest-global-ipos-all-time.asp

Zvonova, E., & Kuznetsov, A. V. (2018). Scenarios of world monetary and financial system development: Opportunities and risks for Russia. *World Economy
and International Relations, 62*(2), 5–16.

CHAPTER 3

Digital India: Cyber Governance, Policing and Diplomacy

Abstract The chapter deals with the aspects of Digital India, the pillars, cyber governance, cyber policing in the States and cyber diplomacy. In terms of cyber policing, a case study of the State of Kerala is taken based on institutional visits and interviews at various organizations, institutions and other agencies. The chapter deals with providing organizational and governance structures, the cyber threats emerging across the country, international cooperation in cyber governance with various countries and international institutions across the world.

Keywords Digital India · Cyber governance · Cyber policing · Cyber diplomacy · Cyber threats

INTRODUCTION

The world is seeing increasing digitization of almost everything. Since the emergence of cyberspace and its global integration, the role of state sovereignty, territorial legitimacy and authority are being questioned. This influences the balance of power, structural and institutional nature of international politics and relations as well. A country like India, which is considered part of a list of countries with greater influence

© The Author(s), under exclusive license to Springer Nature
Singapore Pte Ltd. 2022
R. Reghunadhan, *Cyber Technological Paradigms and Threat Landscape in India*, https://doi.org/10.1007/978-981-16-9128-7_3

on international politics and vice versa, has (and will have) to transform and "technologically update" and digitize in lines with technological advancements in cyberspace.

India is the largest democracy globally, and the International Monetary Fund (IMF) estimates India "to be the fastest-growing" major economy in 2021, reaching a double-digit growth at 11.5% (IMF, 2021). It has among the "largest number of internet users in the world". It reached nearly 700 million users in 2020 and is expected to have a growth rate of 39.14% in 2025 to be close to one billion internet users (Keelery, 2020b). During his presentation at the Cyber Security Conclave 2019 in New Delhi, Dr. V. K. Saraswat, a member of the National Institution for Transforming India (NITI) Aayog,[1] extrapolated that three-fourths of India new users were from rural areas, while mobile video content growth in CAGR was around 83%. About three-fourths of the new users consume data in Indian vernacular languages, which is among the most diverse in any country across the world (Saraswat, 2019).

The Data Security Council of India brought out a National Cyber Security Strategy 2020 that emphasizes on various facets of national cybersecurity. This is divided into three themes and 21 areas. First, securing the national cyberspace: include large-scale digitalization of public services, supply chain security, critical information infrastructure protection, digital payment security, sectoral preparedness, state-level cyber security, security of small and medium businesses or SMB preparedness, advanced technology (including emergent/frontier technologies). Second, strengthen structures, people, processes, capabilities include structures, character, institutions, governance, budget allocation, research, innovation & technology development, capability & skill building, audit & assurance, incident/crisis management and data security & governance. Third, synergizing resources including cooperation & collaboration: include internet infrastructure, standards development, cyber insurance, Brand India, cyber diplomacy, and cybercrime investigation (DSCI, 2020).

During his address to the "Confederation of Indian Industry's Partnership Summit" at Mumbai in 2019, the "Vice President of India, Shri M. Venkaiah Naidu", characterized India as the "bright spot" driving

---

[1] National policy think tank of the Indian Government.

the engine of growth across the globe. It can become a torchbearer for leading the global economy (PIB, 2019). According to him:

> the increasing internet penetration in rural India provides a great opportunity to improve rural India. Connecting all Gram Panchayats with [an] optical fiber network under the Bharat Net project by March 2019 would transform rural India by enabling access to digital transactions and online sales of products, including agri produce via e-Nam. (Ibid.)

India has the largest number of internet users globally and is expected to have a growth rate in the next few years (Keelery, 2020b). Social media users expect to have a voluminous number of users and data being created, competing with China and other advanced economies (Tiezzi, 2020; Keelery, 2020a). Besides being the leading country that drives the global economic engine, a feat driven by an exponential rise in mobile phone users, India is expected to have a humungous rise in digital inclusion in rural areas (Dharna, 2020; Keelery, 2020b). This is an important aspect in bridging the digital divide and blurring if not mitigating the gap between "Bharat vs India". There has been a much higher growth rate in rural areas in terms of data usage than in urban areas. India also became the second largest market of smartphone users and internet users, behind only China (Sanyal, 2020). The Indian e-commerce market is considered to be among the highest growing platforms, much more diverse than most developed nations. It has a huge potential for start-up companies to innovate and grow in the Indian market (*Ecommerce Guide*, 2021; IBEF, 2021a).

The number of Indian users has reached a mammoth scale but has serious implications on the national security aspect. Being a major regional power, India has often been the target of "increasing threats in the form of breaches in data, phishing, trojan horse intrusions, organized cyber-attack, an uncontrolled exploit such as computer worms or virus, malicious software codes, malware attacks, websites being compromised" (Menon, 2018; Reghunadhan, 2018, 38), and other related incidents. The threats and vulnerabilities of cyberspace drive parallel to the digitization activities happening across the world, to which India is not immune. However, an important aspect of this chapter deals with the various facets of Digital India and the smart cities initiative in India, and the transformation the country is going through.

DIGITAL INDIA

The Government of India (GoI) "launched the Digital India programme in July 2015", one year after the Modi Administration approved it in August 2014. According to Prime Minister Narendra Modi, the vision of the programme is "IT (Indian Talent) + IT (Information Technology) = IT (India Tomorrow)" (IBEF, 2014). It provides an important step "into digital empowered society and knowledge economy" (PIB, 2014). This can provide a huge platform for digital governance with India's biggest possibilities and opportunities since its independence in 1947. It is expected to permeate into the country's socio-economic, cultural, demographic and political facets. The level of interconnectedness that the country seems to be moving forward into the era of digitization is something very different from the previous century, especially for India.

An important goal of Digital India is to achieve digital inclusion and connectivity in sectors like finance, banking, health, energy, manu-facturing, agriculture, education, automobile, insurance, defence, phar-maceuticals, electronics, exports-imports, construction, aviation, oil and gas, renewable energy, space industry, tourism, food sector, and railways. The level of "end-user-cum-last-mile connectivity" is something that can inevitably upgrade, strengthen and enhance various sectors in India, and provide an impetus to growth and economic development (MeitY, 2021; Reghunadhan, 2018, 39–40).

The Digital India programme will be India's backbone in its trans-formation into a cashless economy, with further expansion of potential applicability in digital citizenship, elections, democratization, health pass-port and the likes. The "vision areas" include *"Infrastructure as Utility to Every Citizen"*, *"Governance and Services on Demand"*, and *"Digital Empowerment of Citizens"*. Firstly, *Infrastructure as Utility to Every Citizen* emphasizes high-speed internet to all Gram Panchayats, providing digital identity, mobile (digital) banking, easier accessibility to local Common Services Centre (CSC), public cloud facilities and ensuring safe and secure cyberspace.

Secondly, *Governance and Services on Demand* emphasizes on seamless integration of various departments or jurisdictions (single window access), real-time availability of government services, accessibility for citizens to the cloud, digital transformation of government services, e-transfer and cashless transactions in the financial sector, and the utilization of the

Geographic Information System (GIS) to decision-making and development systems. Thirdly, *Digital Empowerment of Citizens* emphasizes digital literacy that is universal, accessibility of digital resources (including documents or certificates) through the cloud, availability of Indian languages-based resources and services, enabling participative governance through collaborative digital platforms, and options for portability of all individual entitlements through cloud technology(ies) (PIB, 2014).

India has a huge e-commerce market and is among the fastest-growing customer base among major economies. There are reportedly 175 million online customers and related transactions, of which 70% are via mobile phone and related accessories (Saraswat, 2019). Nearly half of all travel transactions are estimated to be happening online. In terms of major companies in the digital market and e-commerce, it includes global players like Amazon (through its subsidiary Amazon India) facing local competitors like Flipkart, Infibeam, Paytm, Myntra, IndiaMart, Snapdeal and many others. India's e-commerce market's total worth is expected to reach 99 billion USD in the next three years, "growing at a compound annual growth rate (CAGR) of 27%" (*Ecommerce Guide*, 2021; IBEF, 2021a).

Further, the share of social network user penetration is expected to increase from 50.44% in 2020 to 67.4% of the second-most populous country in 2025. This growth trajectory averages a rate of approximately 61.33%, which is the highest among major economies, behind only China (Keelery, 2020a). It also has the highest concentration of mobile phone users across the democratic world, of which an estimated 46.25% were from the rural population. It provides a huge customer base and accessibility to social networks like Facebook, internet search engines like Google (Dharna, 2020; Keelery, 2020b). In terms of growth in start-ups adopting cyber technologies in India, there has been a growth of 31.25 and 87% in the number of start-ups and investments, respectively, in 2020. NASSCOM is collaborating and supporting various ministries implementing of the Digital India Programme across the nation (PTI, 2021b).

India can be part of the international institutionalization taking place at various parts of the world, and integrate it at the domestic level, providing a huge platform in transforming the conventional mainstream understanding of governance, governability and access to public goods. Thus, it can help achieve what Prime Minister Narendra Modi believes to be "minimum government and maximum governance" (Langa, 2020). In the era of (post-)COVID-19 pandemic, the boundless latitude and the massive digital literacy programme (including telemedicine) across the nation have provided seamless as well as connectivity, at a scale seen hitherto earlier.

PILLARS OF DIGITAL INDIA

The pillars of the Digital India Programme consist of: *"Broadband Highways, Universal Access to Mobile Connectivity, Public Internet Access Programme, e-Governance: Reforming Government through Technology, e-Kranti, Information for All, Electronic Manufacturing, IT for Jobs and Early Harvest Programmes"* (Reghunadhan, 2018, 39–40).

The *Broadband Highways*[2] has the Department of Electronics and Information Technology (DeitY) as the nodal agency in the country. It targets digital inclusivity and connectivity through the infrastructure integration through the National Optical Fiber Network (NOFN) in rural and urban areas. It also focuses on providing internet services and solutions through the upgradation of existing infrastructure, as well as the development of new projects and those already under construction. The initiatives by both the public sector (BSNL, MTNL, etc.) as well as the domestic private sector (Reliance Jio, Airtel, Vodafone, etc.) have been transformative in terms of internet penetration and connectivity.

The nodal department for the National Information Infrastructure (NII) is the "Ministry of Electronics and Information Technology (MeitY)", which focuses on integrating the country's digital network and cloud-related infrastructure. This has been transformative in terms of bureaucratic and governance-related connectivity as well as activities of various departments, viz. e. viz., panchayat-level, taluk-level, district-level, state-level and national-level. The infrastructural components include the National Knowledge Network (NKN), State Wide Area Network (SWAN), National Optical Fiber Network (NOFN), Government User Network (GUN) and MeghRaj Cloud (PMO, 2021; Reghunadhan, 2018).

An Indian subsidiary of the US-based internet-service provider giant Hughes India has partnered with Indian Public Sector Enterprises (PSEs) Telecommunications Consultants India Private Limited (TCIL) and Bharat Broadband Nigam Limited (BBNL) "to provide high-speed

---

[2] Subcomponents involve *"Broadband for All Rural, Broadband for All Urban* and *National Information Infrastructure"*.

satellite connectivity to 5,000 remote gram panchayats" in October 2020. Further, the Union Budget 2020–21 allocated around 5.36 billion USD to India's communication sector, a growth rate of 3.62% from the Union Budget 2019–20 (IBEF, 2020a).

The "Department of Telecommunications (DoT) is the nodal department Universal Access to Mobile Connectivity project". This "focuses on network penetration" and bridging gaps in internet connectivity and accessibility to digital connectivity. Nearly 40,000–55,000 villages will gain access to mobile connectivity, thus providing a huge impetus to achieving universal connectivity and accessibility. The Public Internet Access Programme[3] is implemented. Under the programme, the CSCs would be strengthened and expanded to panchayat-level, focusing on "viable and multi-functional at different end-points for delivering services" (Reghunadhan, 2018).

In 2015, the CSC 2.0 scheme for panchayat-level outreach "under the National Rural Internet Mission (NRIM)". It is expected to cover 250,000 villages with MeitY as the nodal agency and 150,000 post offices with the Department of Posts (DoP) as the nodal agency (MeitY, 2021). This has empowered the district administration through the District e-Governance Society (DeGS),[4] which envisages transaction, capacity building and maximization of efficacy, accessibility and delivery of e-Services (DeitY, 2012; Reghunadhan, 2018).

The e-Governance: Reforming Government through Technology focuses on "re-engineering, simplifying and making government processes efficient and effective in various domains" through e-governance and cyber technological paradigms. This enables streamlining of (digital) procedures, transparency, accountability and accessibility. Further, it focuses on the "integration of services and platforms like Aadhaar, payment gateway, m-Seva platform, sharing of data mandated to facilitate integrated and interoperable service delivery" (Reghunadhan, 2018).[5]

---

[3] Two subcomponents include *common services centres (CSCs) and post offices (as multi-service centres)*.

[4] It is supported by the e-District Manager for e-District National Program Management Unit (NPMU) and District Collector/ District Magistrate. This ensures a dedicated workforce for local level coordination and ensuring delivery of e-District services (DeitY, 2012).

[5] In August 2020, an MoU was signed between the National e-Governance Division (NeGD) under the MeitY with "CSC e-Governance Services India Limited" to initiate

The e-Kranti (Transforming e-Governance for Transforming Governance) is another pillar expected to transform e-governance in the country through the promotion of mobile applications and related accessibility to enhance the efficacy of good governance through cyber technological paradigms in the country[6] (MeitY, 2021). This enhances competitiveness and productivity of governance both qualitatively and quantitatively, improves integration of processes, systems and services, increases connectivity and coordination,[7] increased streamlining and standardization of mechanisms, integration of Indian languages-supported systems, utilization of Next Generation Incubation Scheme (NGIS) for projects, and adherence to strategies, measures and policies of the National Cyber Security Policy (Reghunadhan, 2018).

The *Information for All*[8] is an awareness campaign activity by the GoI, "state governments and related agencies to" provide information

---

delivery of "Unified Mobile Application for New-Age Governance (UMANG)" services, making it "available… through the network of 3.75 lakh CSCs". This is considered to have an effect on enhancing accessibility to government services as well as the income of citizens as well (IBEF, 2020b).

[6] Key principles of e-Kranti are "Transformation and not Translation, Integrated Services and not Individual Services, Government Process Reengineering (GPR), ICT Infrastructure on Demand, Cloud by Default, Mobile First, Fast-Tracking Approvals, Mandating Standards and Protocols, Language Localisation, National Geo-Spatial Information System (NGIS) and Security and Electronic Data Preservation". The various domains include: education (through the provision of internet connectivity at educational institutions, national-level digital literacy programme and options for online courses like the Massive Online Open Courses), healthcare (through provision of online medical consultation, access to medical records, provision of supply of medicines, patient information exchange at pan-India level), agriculture (through "real-time price information, online ordering of inputs", digital exchange of cash or mobile banking), disaster management (through "mobile-based emergency and disaster-related services"), legal system (through Interoperable Criminal Justice System {ICJS} through e-Police, e-Prosecution, e-Courts {which includes the National Judicial Data Grid or NJDG} and, e-Jails/e-Prisons or the NPIP, i.e., National Prisons Information Portal with assistance from Decision Support Systems {DSS}, and Court Management and Case Management System), financial inclusion (through "Micro-ATM program and CSCs/post offices") and cyber security (through National Cyber Security Coordination Centre) (Ecourts Services 2016; MeitY, 2021; NJDG, 2020; NPIP, 2021; Reghunadhan 2018).

[7] It is through cyber technological platforms that provide services related to cloud and mobile platforms and services.

[8] Include initiatives like *Open Data platform for "ministries/departments for use, reuse and redistribute, online hosting of information & documents to facilitate open and easy access to information for citizens"* (Reghunadhan, 2018).

to citizens and various stakeholders through digital platforms in order to facilitate and communicate effective governance strategies.[9]

Meanwhile, Electronic Manufacturing focuses on the promotion of indigenous manufacturing of electronics in India. Recently, through the Make in India initiative (2014), the Production Incentive Scheme (PLI) for Large-scale Electronics Manufacturing and the National Policy on Electronics (NPE) (2019), "Modified Electronics Manufacturing Clusters (EMC 2.0) Scheme" (2020) and Aatma Nirbhar Bharat Abhiyaan[10] (2020) have turned a corner in spurring the "Electronics System Design & Manufacturing (ESDM) sector". The plans to invest 673.2 million USD by the Tata Group in setting up US-based Apple phone component plant (in Tamil Nadu); setting up of 122.5 billion USD from companies such as Foxconn (Taiwan), Pegatron (Taiwan), Rising Star (Taiwan), Samsung (South Korean) and Wistron (Taiwan); and setting up of handset production worth 17.02 billion USD from Indian companies such as Lava, Micromax, Optiemus, Padget Electronics, and UTL Neolyncs in October 2020 will transform India's job market under IT for Jobs as well (IBEF, 2021b; PRSIndia, 2020; Reghunadhan, 2018).

The *IT for Jobs* entails skilling, training, and "availing employment opportunities" in the IT and allied sectors. It emphasizes tapping into the vast potential of human capital in India, which is among the best across the globe. The skilling programme "under the Ministry of Skill Development and Entrepreneurship" for nearly three lakh migrant workers across six districts are more focused on demand-driven through the "centrally sponsored and centrally managed (CSCM) component of the Pradhan Mantri Kaushal Vikas Yojana (PMKVY)" (IBEF,). The North East BPO Promotion Scheme (NEBPS) in Tier-II/III cities would be the focus of the scheme, which intends to tap into Business Process Outsourcing (BPO) through the provision of "reliable internet connectivity and power supply" (MeitY, 2021).

The Early Harvest Programme focuses on short-term implementation requirements and strategies. It comprises of ten projects, namely the IT Platform for Messages project in enabling mass messaging app

---

[9] Government to Citizens (G2C).

[10] National Digital Health Blueprint, e-Vidya, DIKSHA scheme, and National Foundational Literacy and Numeracy Mission.

to increase democratic values, accountability among "elected representatives and Government employees"; Government Greetings to be
e-Greetings project for utilizing MyGov platform to crowdsource e-
designs and e-greetings; Biometric attendance project for covering more
than 40,000 employees from 150 government and related organizations
with e-connectivity; Wi-Fi in All Universities project for by connectivity through National Knowledge Network or NKN[11]; Secure email
within Government project for creating primary internal/departmental
source of communication and focus on upgradation; Standardize Government Email Design project to increase interoperability and coordination;
Public Wi-fi hotspots project in cities with more than one million population, and tourist places of attraction; School Books to be e-Books
project for the conversion of books under the ambit of both MHRD and
MeitY; SMS-based weather information, disaster alerts project through
Mobile Seva Platform under the ambit of MeitY, Ministry of Earth
Sciences (MoES), and Ministry of Home Affairs (MHA); and the
National Portal for Lost & Found children project under the ambit
of MeitY and the Ministry of Women and Child Development entails
"real-time information gathering and sharing" to check crimes against
children, improve response time, (decentralized) participatory response
and awareness (MeitY, 2021; Reghunadhan, 2018).

Cyber Governance in Digital India

The ecosystem of organizational structure that deal with governance of
Digital India programme include: the Prime Minister's Office (PMO);
Ministry of Electronics and Information Technology (MeitY); Ministry of
Agriculture and Farmers' Welfare; Ministry of Commerce and Industry;
Ministry of Communication (MoC); Ministry of Defence (MoD),
Ministry of Corporate Affairs (MCA); Ministry of Education; Ministry of
External Affairs (MEA); Ministry of Finance (MoF); Ministry of Health &
Family Welfare (MoHFW); Ministry of Home Affairs (MHA); Ministry
of Labour & Employment; Ministry of Panchayati Raj; Ministry of Rail-
ways; Ministry of Personnel, Public Grievances and Pensions; Ministry
of Science and Technology (MoST); Ministry of Skill Development &
Entrepreneurship (MSDE); Ministry of Social Justice & Empowerment;

---

[11] Ministry of Human Resource Development (MHRD) is in-charge of implementing
it.

National Association of Software and Service Companies (NASSCOM); National Securities Depository Limited (NSDL); Reserve Bank of India (RBI) and Indian Banks' Association (IBA) (MeitY, 2021).

The Digital India initiative is directly dealt with by the PMO, who heads the "Monitoring Committee on Digital India" (MCDI). Meanwhile, the Minister of Communications and IT heads a "Digital India Advisory Group" (DIAG), and the Cabinet Secretary heads an Apex Committee. Two meetings of the Apex Committee were held in November 2014 and February 2015, respectively. The policy decisions are undertaken by the "Cabinet Committee on Economic Affairs (CCEA)", the DIAG and Apex Committee. These provide inputs and directions to the MCDI, which supervises the implementation.

In terms of financial appraisal/approval, it is under the purview of the "Expenditure Finance Committee (EFC) or Committee on Non-Plan Expenditure (CNE)", which is headed by the Secretary (Expenditure). Some major digital policy initiatives include *Modified Special Incentive Scheme, Electronics Development Fund, National Policy on Software Products* and *New Electronics Policy*. Along with support from the Department of Electronics and Information Technology (DeitY) (with Secretary {DeitY} as standing special invitee), the EFC/CNE recommends to the CCEA the aspects of the implementation of "Mission Mode Projects (MMPs)" and other e-Governance initiatives. The Apex Committee is in charge of harmonizing and integrating "diverse initiative and aspects related to integration of services, end-to-end process re-engineering and service levels of MMPs and other initiatives under the" programme (MeitY, 2021).

The activities related to awareness, sensitization and sharing of best practices will be undertaken by the "Council of Mission Leaders on Digital India, headed by Secretary (DeitY)". The Secretary (DeitY) is also in charge of resolving issues between other bodies, agencies and committees related to the Digital India Programme.[12] A Program Management Information System (PMIS) is constituted to appraise various aspects and

---

[12] A Programme Management Unit, vis-à-vis., the National e-Governance Division (NeGD) has been set up in order to support departments to conceptualize, develop, appraise and implement initiatives under the Digital India Programme. Institutional mechanisms support this "at the State-level, headed by the Chief Minister" (CM) of respective States/UTs, and incidentally, the State-level Apex Committees are headed by Chief Secretaries, respectively (MeitY, 2021).

parameters of MMPs. It acts as a tool to monitor and evaluate "physical, financial and outcome parameters" of MMPs and e-Governance initiatives (PMIS, 2021). The recent initiatives under the Digital India programme include "National Programme on Artificial Intelligence (NPAI), India Enterprise Architecture (IndEA), and Digital North East Vision Document and its Projects" (NeGD, 2021).

Prime Minister's Office

The PMO is the highest decision-making agency in the country that entails policies, strategies and initiatives to deal with threats to the nation's cybersecurity. It coordinates between various Ministries, Departments, government agencies and other stakeholders. Additionally, it entails effective policy-making and intelligence gathering. Further, it takes part in coordinating all enforcement agencies and defence groups related to the country's security. Under the PMO, the major departments, organizations and agencies that are linked with the "implementation of the Digital India programme" include the National Technical Research Organisation (NTRO), National Information Board (NIB), National Cyber Coordination Centre (NCCC) and the Department of Space. The "National Critical Information Infrastructure Protection Centre (NCIIPC)"[13] comes under the ambit of NTRO, while NCCC is run by the "Computer Emergency Response Team-India" (CERT-In) (Dilipraj & Reghunadhan, 2018, 121; PMO, 2021).

Ministry of Electronics and Information Technology

In terms of being the major nodal agency, MeitY is in charge of "transforming the nation into a digital superpower" while acting "as an enabler" in "empowering the citizens" and in providing avenues for the perpetuation of digital infrastructure, governance, know-how and literacy. The departments, organizations and agencies that are linked with the implementation of the Digital India programme include the "Department of Electronics and Information Technology (DeitY), Centre for Development of Advanced Computing (CDAC)", Commons Services Centre (CSC), "Controller of Certifying Authorities (CCA), Cyber Appellate

---

[13] It focuses on the identification and protection of Critical Information Infrastructure (CII).

Tribunal (CyAT)", Cyber Swachhta Kendra (CSK)[14]; Education and Research Network (ERNET), "Indian Computer Emergency Response Team" (ICERT) or CERT-In, National Centre of Geo-Informatics (NCOG), "National Informatics Centre (NIC), National Institute of Electronics and Information Technology (NIELIT), National Internet Exchange of India (NIXI)", Standardisation Testing and Quality Certification (STQC), Unique Identification Authority of India (UIDAI). MeitY[15] directly supervises and coordinates agencies that provide and deal with the information sharing ecosystem, cyber defensive and cyber offensive strategies. It also functions closely on the prevention and mitigation of disasters and related crises (MeitY, 2021; Dilipraj & Reghunadhan, 2018, 122–123).

There are digital literacy programmes like Pradhan Mantri Gramin Digital Saksharta Abhiyan (PMGDISHA), National Digital Literacy Mission, etc. The Ministry focuses on terms of transforming and transitioning the economy into "a digitally empowered society and knowledge economy", one which enables the creation of a "[f]aceless, [p]aperless, [c]ashless" society (MeitY, 2021).

Ministry of Agriculture, & Farmers' Welfare

Under the "Ministry of Agriculture & Farmers' Welfare, the Department of Agriculture Cooperation and Farmers Welfare (DAC&FW)" coordinates with state-level organizations and agencies to implement schemes in the agricultural and allied sector. The Small Farmers Agribusiness Consortium (SFAC), an autonomous society, focuses on small and marginal farmers through support to agribusiness[16] incomes and the implementation of the National Agriculture Market Electronic Trading (e-NAM) platform. The "Indian Council of Agricultural Research

---

[14] Also known as Botnet Cleaning and Malware Analysis Centre.

[15] It consists of various divisions to deal with cyber technological paradigms and related governance in the country. These include Electronics System Design & Manufacturing, Infrastructure & Governance, IT/ITeS, Economic Planning, Digital Economy & Digital Payment Division, Cyber Security Division, "International Co-operation, Trade and Investment, Human-Centred Computing", e-Governance, "Emerging Technologies Division", Cyber Laws & E-Security, Research & Development (MeitY, 2021).

[16] Farmer Business Companies (FBC) and/or Farmer Producer Organization (FPO).

(ICAR)", also an autonomous society under the Department of Agricultural Research and Education", coordinates and manages R&D in the agricultural sector (DeitY, 2012; MeitY, 2021; MoA&FW, 2021). The ministry has initiatives and schemes like the National Soil Health Card, e-NAM platform, etc.

Ministry of Commerce and Industry

Under the Ministry of Commerce and Industry, the Directorate General of Supplies and Goods (DGS&D) focuses in developing an "online e-procurement portal" among ministries, governmental departments, and other related agencies. The "Department of Industrial Policy & Promotion (DIPP)" is the nodal department for the implementation of Industrial Policy, and thus an important nodal point in determining the allocation, integration and implementation of the Digital India programme, as well as "Make in India" and "Skills India" across the industrial sector in the country. This is important in the creation of an "Export Promotion Mission", enable the provision of digital refund to exporters, technology-driven solutions for "Ease of Doing Business" and Common Digital Platform (CDP) for e-Certificate of Origin (e-COO) under Aatma Nirbhar Bharat Abhiyan,[17] for the MSME sector, the Market Access Initiative (MAI), digital registration-cum-membership certificates (RCMCs), Engineering Exports Promotion Council of India (EEPC India) digital interface, the digital marketing at district-level, "Metals and Minerals Trading Corporation of India enterprise resource planning" (MMTC ERP) system upgradation and/or migration, BHIM app, e-commerce and options for e-certificate through the Directorate General of Foreign Trade Common Digital platform since April 2020 (Department of Commerce, 2021, 1–174; DIPP, 2021; MeitY, 2021; GoI, 2020).

Ministry of Communication

Under the MoC, the Department of Telecommunication (DoT) as a nodal point focuses on "providing services, issuance of guidelines,

---

[17] AatmaNirbhar Bharat Abhiyan was launched in May 2020. Under its ambit, the new Directorate General of Foreign Trade (DGFT) platform uses IT tools like AI, data analytics, API-based interfaces, etc.

and taking necessary action with regard to infrastructure development for supporting the cyberspace framework in the country" (Dilipraj & Ramnath, 2018, 122). Under the DoT various organizations, companies and autonomous bodies deal with cyber technological paradigms and threats that include the "Digital Communications Commission (DCC), Telecom Regulatory Authority of India (TRAI)",[18] Wireless Planning & Finance, "National Institute of Communication Finance (NICF)", Telecommunication Engineering Centre (TEC), National Operations Control Centre, National Centre for Communication Security, "National Telecom Institute for Policy Research, Innovation and Training (NTIPRIT)", Telecom Enforcement Resource and Monitoring (TERM), "Telecom Disputes Settlement and Appellate Tribunal (TDSAT)", Director General Telecom Head Quarters (DGT-HQ), "Universal Service Obligation Fund (USOF)", Wireless Monitoring Organisation (WMO), Bharat Sanchar Nigam Limited (BSNL), Telecommunications Consultants India Limited (TCIL), Indian Telephone Industries Limited (ITILD), "Mahanagar Telephone Nigam Limited (MTNL), Centre for Development of Telematics" (C-DOT), Bharat Broadband Network Limited (BBNL), Central Public Procurement Portal (CPPP), "Controller General of Communication Accounts (CGCA)", and Office of Controllers of Communication Accounts (CCA) (Dilipraj & Ramnath, 2018, 123–124; DoT, 2021; TDSAT, 2021; TRAI, 2021).

Ministry of Defence

Under the MoD, the major focus is on the "defence, response and resilience of the nation in case of an imminent cyber-related attack". The Defence forces have respective CERTs to deal and resolve cyber-related threats and attacks, coordinates with the MoD-CERT. The issues related to fake news, social media and metadata of internet traffic are dealt with by the "Defence Research and Development Organisation (DRDO)", while the aspects related to cyber intelligence are under the ambit of the Defence Intelligence Agency (DIA). A Cyber Unit has been institutionalized to deal with defensive and attribution-related aspects and comprises of personnel from MoD, MEA and Defence forces. The Head Quarters (HQ) of the Integrated Defence Staff (IDS) comes under the

---

[18] TRAI has utilized the blockchain technology to control spam SMS traffic (*ET Bureau*, 2021).

aegis of the Chairman of Chiefs of Staff Committee and houses three important organizations, namely "Defence Information Assurance and Research Agency (DIARA), Defence Intelligence Agency (DIA)" and the upcoming Defence Cyber Agency (DCA), Defence Research and Development Organisation (DRDO) (managing the Network Traffic and Analysis or NETRA programme) (DRDO, 2020; Dilipraj & Ramnath, 2017; *ET Online*, 2017; Paganini, 2014).

Ministry of Corporate Affairs

Under the MCA focuses on administering laws, regulations and strategies for the corporate sector under the purview of the Digital India programme. The Institute of Chartered Accountants of India Digital Learning Hub provides an integrated Learning Management System (LMS) focusing on the emergence of the country's knowledge economy. It is managed by the Digital Re-Engineering & Learning Directorate and is facilitated by an ICAI committee or department (Institute of Chartered Accountants of India, 2021; MeitY, 2021). The "Institute of Company Secretaries of India" (ICSI) has various initiatives including facilities like e-Learning, e-Library, the e-Management Skill Orientation Programme (E-MSOP), conducting Computer-Based Examination (CBE), the Student Member Application Software Hosting (SMASH) app for integrated access to services of ICSI and virtual collaboration platforms (ICSI, 2021). The Institute of Cost Accountants of India has been part of initiatives for skilling and encouraging Digital India, Startup India, Skill India programmes across the country (Institute of Cost Accountants of India, 2021; MeitY, 2021).

Ministry of Home Affairs

Under MHA, the role centres on providing security guidelines, and in assisting in efforts to deal with attacks that have internal security-related implications for the country.[19] Its role is important in sensitizing other ministries, departments, agencies and organizations in dealing with threats to critical national and/or information infrastructure. It ensues on strengthening existing security measures and mechanisms, increasing

---

[19] The Cyber and Information Security (C&IS) Division consists of Coordination Wing, Cyber Crime Wing, Information Security, Monitoring Unit and I4C.

interoperability, cyber policing and storage. Meanwhile, it focuses on developing further to plug gaps and even mitigate vulnerability for potential cyber-related attacks inimical to Digital India. It helps secure the hard infrastructure (tangible and/or built infrastructure), strengthens security measures (resilience against attribution-based attacks) and increases inter-departmental coordination (interoperability) and awareness in dealing with threats. The departments, organizations, agencies and schemes that are linked with the implementation of the Digital India programme include National Intelligence Grid (NATGRID), "Indian Cyber Crime Coordination Centre" (ICCCC or I4C) Scheme,[20] "Crime and Criminal Tracking Network and Systems (CCTNS)", "Cybercrime Prevention Against Women and Children Scheme" (CCPWC),[21] Common Integrated Police Application (CIPA) and Security Alert System (SAS) (Dilipraj & Reghunadhan, 2017; Dilipraj & Reghunadhan, 2018, 124; Department of Kerala Police, 2021; MHA, 2021).

Ministry of External Affairs

MEA is the nodal agency in establishing cyber diplomatic overtures and cooperation with countries (through bilateral engagement), and internationally (through multilateral and inter-governmental engagement).[22] The focus of the Ministry has been on initiating the "digital diplomacy footprint" through digital initiatives like the Global Pravasi Rishta portal and app to enable "effective communication" between the Ministry, Indian missions and diaspora; the website Indbiz.gov.in to

---

[20] The I4C Scheme coordinates with the Cybercrime Ecosystem Management Unit, "National Cybercrime Forensic Laboratory" (NCFL) Ecosystem, National Cybercrime Reporting Portal, National Cybercrime Research and Innovation Centre, "National Cybercrime Threat Analysis Unit" (TAU), National Cybercrime Training Centre (NCTC), and "Platform for Joint Cybercrime Investigation Team" (MHA, 2021).

[21] Five components of CCPWC Scheme are *Awareness Creation Unit, Capacity Building Unit, Forensic Unit, Online Cybercrime Reporting Unit* (citizen-centric portal maintained under CCTNS) and *R&D Unit*. CyberDost launched for cybercrime awareness through Twitter, setting up of CoE for R&D in the domain of Cyber Crime Prevention & Control, workshops and courses for cybercrime investigation, hiring of cyber forensic consultants, setting up of Cyber Forensic training Labs (MHA, 2021).

[22] There is Cyber Diplomacy Division, and the E-Governance & Information Technology Division (MEA, 2021a).

show sectoral and state-wise trends and increase foreign capital investment; MEA Performance Smartboard to monitor the performance of the Ministry; MADAD portal that emphasized on actions to file consular, forward and handle "complaints, improve tracking and redressal, escalate unresolved cases"; MEA App is increasing digital media base; E-Audit Portal; "Foreign Service Institute Alumni Portal; Revamped Knowledge India Programme; Pravasi Bharatiya Divas Portals; Diplomatic Identity Card Registration and Issuance System"; SEWA-Indian Consular Services System to apply for passport and visa services; and the Passport Seva App to digitally interconnect the Passport Seva Kendra (PSK). Targeting "paperless governance", the Passport Seva Program introduced the DigiLocker platform to digitally connect with any applicant, with services provided through "paper-less mode" across the country. The e-Passport initiative under the upcoming "Passport Seva Programme 2.0... [will use] emerging technologies such as Artificial Intelligence (AI), Machine Learning, Chat-Bot, Analytics, Robotic Process Automation (RPA)" (MEA, 2021b; MEA, 2020b; MEA, 2020c).

Ministry of Education

Under the MoE, the focus is on two categories of schemes: University and Higher Education, and Technical Educations. The main agencies that support and implement the Digital India Programme under the MoE[23] are the "National University of Educational Planning and Administration (NUEPA),[24] and the National Council of Education Research and Training" (NCERT)[25] (MoE, 2021).

[23] This includes the Indian National Digital Library in Engineering, the Innovation Universities Aiming at World Class Standards, the National Initiative for Design Innovation, the National Programme for Technology Enhanced Learning (NPTEL), the National Research Professorship (NRP), the Rashtriya Ucchatar Shiksha Abhiyan (RUSA), the schemes under the University Grants Commission (UGC) and other statutory and autonomous organizations, the Science & Technology (INDEST-AICTE) Consortium, Technology Development Mission, and the initiative to establish the Indian Institutes of Information Technology (IIITs).

[24] A "deemed to be university", focusing on capacity building, planning and management of education as well as R&D.

[25] An "autonomous body" having an advisory role with regard to the improvement of quality of education in the country.

Ministry of Finance

MoF is an important nodal point for the Digital India Programme, under which the Department of Finance Services play an important role.[26] There are institutions like the National Skill Development Corporation, Securities Appellate Tribunal and "Securities and Exchange Board of India" supporting and implementing the Digital India programme. MoF has focused on[27] aspects like increasing digitization-related initiatives,[28] encouraging of digital transactions,[29] accountability and transparency in the digital economy, digital literacy, skilling through digital, introducing digital mode in electoral funding, tax collection[30] and the promotion of the digital economy[31] (MoF, 2021; MoF, 2017).

Ministry of Personnel, Public Grievances and Pensions

Under MoPPGP, there are various programmes and systems like the National Digital Crime Resources and Training Centre (NDCRTC), the Integrated Government Online Training Programme (iGOT), "State Category Training, Programme (SCTP); Trainer Development Programme (TDP); Intensive Training Programme; Induction Training Programme; Comprehensive Online Modified Modules for Induction Training (COMMIT)"; Augmentation of the Capacity of Training Institutions (ACTI), "Smart Performance Appraisal Report Recording Window (SPARROW)", Vigilance Information System (VIS), Web-based cadre management of the Central Secretariat Service (CSS), Institute

---

[26] Besides, it also includes various divisions like the Administration & Coordination Division, Budget Division, Infrastructure Policy & Finance Division, Bilateral Cooperation Division, Economic Division, Financial Markets Division, Investment Division, Integrated Finance Division and International Economic Relations Division.

[27] 2017–18 budget speech of the Minister of Finance.

[28] Initiatives like BharatNet, DigiGaoin.

[29] Through the usage of apps like BHIM (including "Referral Bonus Scheme" for individual-specific purposes, and "Cashback Scheme" for merchant-specific purposes); Aadhaar Pay for merchants.

[30] Revenue, Accountability, Probity, Information and Digitisation (RAPID).

[31] Through "m-POS, micro-ATM standards version 1.5.1, Finger Print Readers/Scanners and Iris Scanners".

of Secretariat Training and Management (ISTM), "Training Management Information System (TMIS), and Recognized Trainer Development Programme on Direct Trainer Skill and Design of Training" (RTDP-DTS & DoT). MoPPGP focuses on training and (re)skilling the bureaucratic structure in the country (MoPPGP, 2017; MoPPGP, 2020).

Ministry of Health & Family Welfare

Under the MoHFW, the National Health Mission (NHM) and the "National Institute of Health and Family Welfare" are the nodal agencies implementing the Digital India Programme. The "National Digital Health Mission (NDHM)" was launched to focus on digitizing healthcare in the country, and was set out in the "National Digital Health Blueprint (NDHB)". NDHM is under the ambit of the National Health Authority (NHA), implementing under the "Ayushman Bharat-Pradhan Mantri Jan Arogya Yojana (AB-PMJAY)". The digital systems include Health UD (in selected Union Territories), Digi-Doctor, Health Facility Register (HFR), "Personal Health Records (PHR) and Electronic Medical Records (EMR)". MoHRW focuses on creating a "strong public digital infrastructure... to expand the reach of digital health" and the integration of emergent technologies for digital management (MoHFW, 2021; Rastogi, 2020).

Ministry of Labour & Employment

Under the MoL&E, various initiatives and services have been launched, including the "Santusht" portal, Implementation Monitoring Cell (IMC) and the Centralized Public Grievance Redressal and Monitoring System (CPGRAM). These initiatives by the MoL&E entail bringing "further transparency, accountability, effective delivery of public services and implementation of policies, schemes of [the] Ministry" under the Digital India Programme (*ET Government*, 2020; MoL&E, 2021).

Ministry of Panchayati Raj

Under the MoPR, the focus has been to "strengthen e-Governance" in the Panchayati Raj Institutions (PRIs) within India. The MoPR has emphasized the Gram Panchayat Development Plan (GPDP) and the "Rashtriya Gram Swaraj Abhiyan (RGSA)" through the launch of

"Deen Dayal Upadhyay Panchayat Sashaktikaran Puraskar (DDUPSP), Nanaji Deshmukh Rashtriya Gaurav Gram Sabha Puraskar (NDRGGSP)", Panchayat Decision Support System (PDSS), Gram Panchayat Development Plan (GPDP), National Assets Directory (NAD),[32] e-GramSwaraj web portal, PRIASoft-PFMS integration[33] and Gram Manchitra. This focuses on increasing decentralized accessibility for the citizens and to increase transparency and accountability in "planning, progress reporting and work-based accounting" (MoPR, 2021; MoPR, 2019).

Ministry of Railways

Under the "Ministry of Railways", the "Centre for Railway Information Systems (CRIS)" focuses on digitizing railway operations and monitoring, on real-time tracking analysis, providing data centre and cloud facilities, enterprise-level integrated solutions, smart mobile app facilities for support and other related services. It is the nodal agency for the integration and modernization of IT solutions and emergent technologies in the railways. The adoption, integration and scalability of cloud technologies are part of the "Cloud First" policy, and includes the "Infrastructure as a Service (IaaS), Platform as a Service (PaaS), Network as a Service (NaaS), and Backup as a Service (BaaS)". Further, the agency supports apps for features like unreserved ticketing system (UTS), "passenger reservation system (PRS), national train enquiry system (NTES)", and provides safety support to the railway police force (RPF). The apps include HRMS, Rail Saarthi, RESS, Rail Connect, TDMS, IREPS, Chalak Dal, Rail Madad, Rail Sugam, ICMS and Rail Saver (CRIS, 2021).

Ministry of Science and Technology

Under the "Ministry of Science and Technology" (MoST), the "Department of Science & Technology" (DST) is the nodal point of implementing the Digital India programme and is involved in the development of multiple technologies complementing and supplementing the programme. MoST has various technology focus programs like Big

---

[32] "NAD is one of the software applications envisaged as part of Phase II of the e-Panchayat MMP".

[33] Panchayati Raj Institutions Accounting Software (PRIASoft).

Data Initiative, National Spatial Data Infrastructure, Natural Resource Data Management System, National Supercomputing Mission, Cognitive Science Research Initiative (CSRI), etc. The focus of technology development programs has been evident through various technology projects that include convergence with the likes of "Artificial intelligence, Cyber Security, Supercomputing, Big Data, Machine Learning, Robotics & automation and so on". It supports the promotion of futuristic studies and emergent areas through financial and economic support, innovation and technology start-up support, coordination and integration. MoST has been supporting research in areas like an automated surveillance system, cloud computing platforms, consumer broadband labels, development of renewables to support smart cities, distributed computation capabilities, embedded systems and related technology, infrastructural developments in academic institutions, Internet of Things Space (including cyber systems and information assurance), IP chip design, Machine learning for data mining, technological advancements to increase reliability and secure systems against Trojan, open IE inference, process documentation in government portals, scalability through OS development, support for R&D in Electronics and IT, etc., development of ultra-dense 5G cognitive radio network and wireless sensor networks (DoST, 2021). The National Mission on Interdisciplinary Cyber-Physical Systems (NM-ICPS)[34] has also been launched by MoST, which will establish 25 Technology Innovation Hubs (TIHs) on major academic and research institutions in the country.

---

[34] The mission management includes a "Mission Governing Board (MGB) supported by an Inter-Ministerial Coordination Committee (IMCC) and Scientific Advisory Committee (SAC). In addition, there will be Subject Expert Committees, Sectoral Committees, Cluster Committees and International Advisory Committees to look into the specific requirements of Mission Implementation. Each of these Committees will have subject experts, academicians and industry partners will be involved in implementation of projects. Mission will create a database with initial target of 100 experts to be involved in implementation of projects... The proposed Mission would act as an engine of growth that would benefit national initiatives in health, education, energy, environment, agriculture, strategic cum security, and industrial sectors, Industry 4.0, SMART Cities, Sustainable Development Goals (SDGs) etc." (NM-ICPS, 2021).

Ministry of Skill Development & Entrepreneurship

Under the MSDE, the "National Skill Development Corporation" is the nodal agency for the promotion and implementation of skill development in line with the emerging digitization challenges in India. It focuses on the implementation of Skill India through various schemes and initiatives like the "Pradhan Mantri Kaushal Vikas Yojana (PMKVY)", the establishment of "Pradhan Mantri Kaushal Kendras" (PMKKs) or Model Training Centres (MTCs), "India International Skill Centre (IISC) Network", and "Technical Intern Training Program (TITP)". Further, it provides services such as SMART, SDMS, "Takshashila (National Portal for Training and Assessors), Kaushal Mart (Skilling Resource Marketplace), Kaushal e-Pustakalaya", and collaboration with DeAsra Foundation for skilling and entrepreneurship support (NSDC, 2021). Under the "Ministry of Social Justice & Empowerment", the "Department of Empowerment of Persons with Disabilities (DEPWD)", also known as Divyangjan, is focused on skilling and providing support to Persons with Disabilities. This is implemented through the Information and Communication Eco-System Accessibility provision under the Accessible India Campaign, R&D on "assistive technology and product development devices", support, modernization and capacity augmentation of Braille Presses, and the institutionalization of Unique Disability ID (UDID) for Persons with Disabilities project ranging from village to national level (DEPWD, 2021).

Other Agencies

Besides the ministries, there are some other bodies and agencies that are nodal agencies for the implementation of the Digital India programme. These include the "National Association of Software and Service Companies (NASSCOM), National Payments Corporation of India (NPCI), NSDL Database Management Limited (NDML) and the National Securities Depository Limited (NSDL)". NASSCOM is an industrial body in India, brought out the Deep Tech Club (DTC) 2.0 to support more than 1000 Indian technology start-ups in the domain of cyber technologies and other frontier technologies. This included "artificial intelligence, machine learning (ML), AR, VR (augmented and virtual reality), IoT (Internet of Things), Robotics, Blockchain, NLP (natural language processing)", etc., through three phases (PTI, 2021b). NASSCOM is collaborating with Microsoft, bringing forth a new AI Gamechangers programme, and is expected to organize the "Xperience AI Summit" in July 2021. The integration of AI can provide a huge impetus to unlocking India to become a major stakeholder in the global innovation hub, with an estimated economic value of 500 billion USD of Indian GDP by 2025 (IANS, 2021b). NASSCOM is also partnering with the US-based Cisco[35] and the All India Council for Technical Education (AICTE) to offer internship and skilling, in two phases for the year 2021 (IANS, 2021a).

Under the "National Securities Depository Limited (NSDL), the NSDL Database Management Limited (NDML)" is linked with implementing the Digital India Programme. NDML (along with support from NSDL) provides "technological expertise, physical network and management experience" to projects of "national importance". It provides operational support and solutions in the form of "Information and Communication Technology interventions" to streamline government services and "enhance service delivery standards". The projects implemented include Special Economic Zones (SEZ) Online, the National Payment Services Platform,[36] the National Skills Registry (NSR),[37] the KYC Registration Agency (KRA),[38] NSDL National Insurance Repository (NIR),[39] and the National Academic Depository (NAD).[40] NSDL is involved in other projects like "Tax Information Network (TIN)", Central Recordkeeping Agency (CRA) and Aadhaar (NDML, 2021). Another organization was the creation of a collaboration between the Reserve Bank of India (RBI) and the "Indian Banks' Association (IBA)", known as the "National Payments Corporation of India (NPCI)" (NPCI, 2021).

NPCI deals with the "operating retail payments and settlement systems" in the country". It has "ten core promoter banks", and is more "is focused on bringing innovations in the retail payment systems

---

[35] Nasscom FutureSkills Prime platform of NASSCOM, and NetAcad platform of Cisco.

[36] In collaboration with DeitY.

[37] In collaboration with NASSCOM.

[38] Along with the "Securities and Exchange Board of India (SEBI)".

[39] Established by NDML with approval from the "Insurance Regulatory and Development Authority of India (IRDAI)".

[40] It was established by the "Ministry of Human Resources Development" to achieve digital enablement of the Education Records.

through the use of technology for achieving greater efficiency in operations and widening the reach of payment systems" The initiatives include Unified Payments Interface (UPI), RuPay (including RuPay Contactless, and Bharat E-commerce Payment Gateway), Bharat Interface for Money (BHIM), National Automated Clearing House (NACH), "Immediate Payment Service (IMPS), National Electronic Toll Collection (NETC)" FASTag,[41] Bharat BillPay, *99# (USSD-based m-banking service), Cheque Truncation System (CTS), "National Financial Switch (NFS), Aadhaar Enabled Payment System (AePS)", BHIM Aadhaar payment service, and Certifications Zone to provide hassle-free self-certification web-based tools and services. These initiatives have increased accessibility, interoperability, cost-effectiveness, integration, effective management of complaints and disputes, standardization and scalability. Further, NPCI focuses on dealing with the emerging and complex cyber threat landscape related to fintech (NPCI, 2021).

Cyber Threat Landscape in Digital India

There has been an increasing incidence of cyber-related attacks like phishing, spamming, malware attacks, digital frauds on individuals, organizations as well as institutions. The major issue with the crimes in cyberspace is the dismal number of cases reported often due to fear, lack of awareness among the victim and/or investigators as well as adjudicators and the inherent loopholes that exist in legislations, policies and practices. The incidents related to phishing, unauthorized network scanning or proving, attacks from viruses and malicious codes have been increasing in India. Unauthorized network scanning and probing incidents saw a huge jump of 264% in 2005, increasing from 40 incidents (2005) to 305,276 incidents (2019) being handled by CERT-In. This amounted to CERT-In handling approximately three unauthorized network scanning and probing incidents per month (as of 2005) to 836 incidents per day in 2019 (Table 3.1).

In the first decade of the twenty-first century, the incidents averaged approximately 214 incidents annually, with an average yearly growth rate of 109.45%. In the second decade, the incidents averaged approximately 50,822 incidents annually, with an average yearly growth rate of 454.02%.

---

[41] NETC FASTag employs Radio Frequency Distribution (RFID).

Table 3.1 Cyber threat related to phishing, network scanning, probing and virus or malicious code dealt by CERT-In (2005–19)

Year	Number of incidents			Growth rate		
	P	UNS/P	V/MC	P	UNS/P	V/MC
2005	101	40	95	3266.67	263.64	1800
2006	339	177	19	235.64	342.50	−80
2007	392	223	358	15.63	25.99	1784.21
2008	604	265	408	54.08	18.83	13.97
2009	374	303	596	−38.08	14.34	46.08
2010	508	277	2817	35.83	−8.58	372.65
2011	674	1748	2765	32.68	531.05	−1.85
2012	887	2866	3149	31.60	63.96	13.89
2013	955	3239	4160	7.67	13.02	32.11
2014	1122	3317	4307	17.49	2.41	3.53
2015	534	3673	9830	−52.41	10.73	128.23
2016	757	416	13,371	41.76	−88.67	36.02
2017	552	9383	9750	−27.08	2155.53	−27.08
2018	454	127,481	61,055	−17.75	1258.64	526.21
2019	472	305,276	62,163	3.97	139.47	1.82

Source Compiled by the Author; *Note*: P = Phishing; UNS/P = Unauthorized Network Scanning/Probing; V/MC = Virus/Malicious Code

The increase in incidents related to viruses and malicious codes were next in line, with a huge jump of 1800% in 2005. It increased from 95 incidents (2005) to 62,163 incidents (2019) being handled by CERT-In. This amounted to CERT-In handling approximately eight viruses and malicious code incidents per month (as of 2005) to approximately 170 incidents per day in 2019. In the first decade of the twenty-first century, the incidents averaged approximately 716 incidents annually, with an average yearly growth rate of 109.45%. In the second decade, the incidents averaged 18,950 incidents annually, with an average yearly growth rate of approximately 79.21%. The increase in incidents related to phishing attacks saw a huge jump of nearly 3267% in 2005. It increased from 101 incidents (2005) to 472 incidents (2019) being handled by CERT-In. This amounted to CERT-In handling approximately eight phishing incidents per month (as of 2005) to approximately 39 phishing incidents per month in 2019. In the first decade of the twenty-first century, the incidents averaged approximately 386 incidents annually, with an average yearly growth rate of 594.96%. In the second decade, the incidents averaged approximately 712 incidents annually, with an average yearly growth rate of 4.21% (Table 3.1).

In the first decade of the twenty-first century, it averaged 3,316 incidents being dealt by CERT-In, an average growth rate of 266.7% annually. In the second decade, it averaged 110, 375 cybersecurity-related incidents being dealt by CERT-In, an average growth rate of 81% annually. The highest growth rate during this period was 1004.35% in 2005, while (negative) growth rate was seen a decade later at -62.1% in 2015. The years 2015–17 showed the lowest growth rate in incidents, with an average (negative) growth rate of -18.3%. This exponentially increased in 2018, which saw the second highest growth rate in the twenty-first century, which reached 208,456 incidents in one year. An important rise in 2018 and further in 2019 can be attributed to increasing technological wherewithal of the organizations and personnel in dealing with attacks as well as increasing vulnerabilities due to interconnectedness in the country under the Digital India initiative (Fig. 3.1).

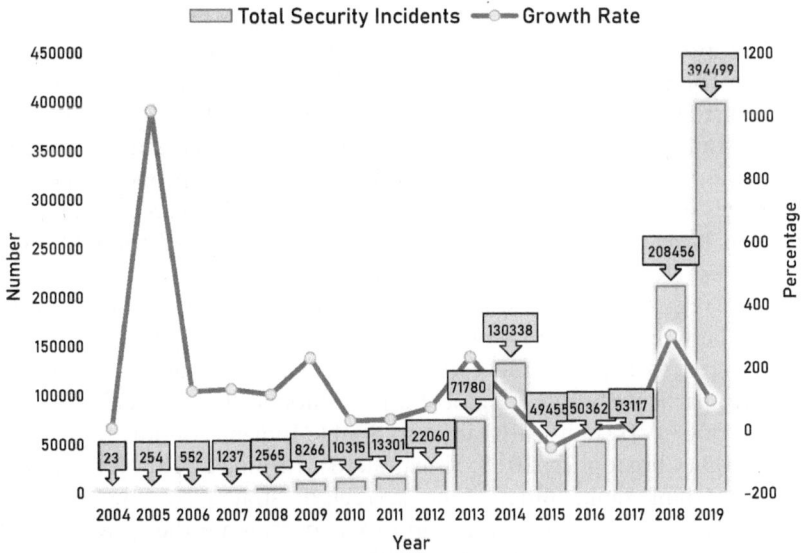

Fig. 3.1 Cybersecurity incidents dealt by CERT-In during 2004–19 (*Source* Compiled by the Author)

Artificial Intelligence in India

India is beginning to take initial steps in AI, one which is taken seriously by the government over the last few years. The Prime Minister of India unveiled the country's first AI research institute in the state of Maharashtra, called Wadhawani AI at Mumbai, in the State of Maharashtra in 2018. But overall, the development of AI in India is relatively considered below par. At a modest 480 million USD in 2018 (includes machine learning and IoT), the allocation, in comparison to other major powers, is still unviable for India, when the country projects itself as a Great Power or a regional power. An analysis report in 2018 revealed that nearly 15% of the research in India is done by the industry, which is only one-fifth of the research being done in the universities. Again, nearly 70% of the research done in India is by companies headquartered outside India like Google, Facebook, IBM and the like (Christopher, 2018; Sinha, 2018).

According to *Global AI Talent Report 2019*, India comprised of only 1.72% of the total number of PhD scholars and academicians in the world, with impactful research work being limited to just 0.22%. India was way behind leading countries like the US, China, UK, Australia and Canada. As an important aspect dealt within the survey, India was listed among the top ten countries with a gender ratio at 17%. But this was less than China and Australia (22%), the US (20%), Switzerland (19%), UK and Italy (18%). India was also categorized as among anchored countries, wherein there was "less talent inflow and less talent outflow, as a proportion of the country's overall talent pool". In terms of "impactful research", India was close to Israel at 29%, while China (90%), Italy (86%) and the US (84%) led the list (Karmanov & Hudson, 2018).

With regard to the Research and Development (R&D) in AI, India spends between 0.5 to 0.7% of its Gross Domestic Product (GDP), which is expected to triple in the coming years. This is a huge challenge to achieve, particularly due to global and economic slowdown due to the impact of COVID, and with target to double Indian economy to five trillion USD (Christopher, 2018; Sinha, 2018). In June 2018, NITI Aayog, the GoI think tank, initiated the *#AIForAll* strategy or the *National Strategy on Artificial Intelligence* in order to promote the growth of AI ecosystem and institutions in India. This is in line with the GoI policy of *"Sabka Saath Sabka Vikas"*, which means "Together, for everyone's growth" (NITI Aayog, 2020).

MeitY formed four committees in looking at furthering policy-level prospects of AI in the country, dealing with aspects of citizen-centric services, data platforms, (re-)skilling, R&D, legal regulatory and cyber-security (*Rajya Sabha TV*, 2018a). NITI Aayog has identified five fields to focus on implementing AI: "healthcare, agriculture, education, smart cities and infrastructure, smart mobility and transportation" (Hebbar, 2018). The think tank is collaborating with technology giants like US-based corporations like Google, Microsoft, Intel, International Business Machines Corporation (known as IBM), Switzerland-based Asea Brown Boveri's (known as ABB) Indian subsidiary called ABB India, and Germany-based Systems Applications and Products in Data Processing Societas Europaea (known as SAP SE) (Ibid.).

In 2020, the GoI launched the *National AI Portal of India*, and a National Program for the country's youth (with the motto "*Responsible AI for Youth*"). The portal was a joint venture between MeitY, NeGD and NASSCOM. It focuses on providing updates on AI-related "developments in India, sharing of resources like articles, startups, investment funds in AI, resources, companies and educational institutions... documents, case studies, research reports... [and] job roles related to AI". Meanwhile, the programme for youth targets the potential of "students of classes 8 to 12 from Central and State government-run schools (including KVS, NVS, JNV) from across the country". The selected students are provided further training through residential boot camps or online sessions" (MeitY, 2020).

There are many major educational institutions providing courses in AI, which include the Indian Institute of Science (IISc), Indian Institute of Technology (IITs), Indian Institutes of Information Technology (IIITs), Chandigarh University, Indraprastha Institute of Information Technology, Great Lakes International University, SRM Institute of Science and Technology, Vellore Institute of Technology, DY Patil International University, University of Petroleum and Energy Studies, Jain University, GH Raisoni College, Sri Ramaswamy Memorial Institute of Science and Technology (SRM) University, Sharda University, Quantum University, Sage University, Shri Vaishnav Vidyapeeth Vishwavidyalaya, Dehradun Institute of Technology (DIT) University, Lovely Professional University, JK Lakshmipat University, Galgotias University, CT University, etc., (Srivastava, 2019).

The AI-related jobs in India are expected to grow from at a rate of 31.6% CAGR from 2018 to 2023, with major companies involved in providing huge impetus to skilled personnel in India. The major companies in India include Microsoft, Facebook, Amazon, Apple, Nvidia, Siemens Technology and Services Private Limited, OpenAI, H20.ai, Synopsys, Citi, Aptiv, Ubisoft, Grammarly, x.ai, DataRobot, Narrative Science, Clarifai, AlphaSense, Tempus, etc., (Kumar, 2021a, 2021b, 2021c). This mainly saw the rise of new type of professions and jobs like machine learning engineer, machine learning scientist, computer vision engineer, AI software engineer, solution architect, algorithm developer, etc., (Kumar, 2019). Business analytics has become an important requirement for successful business in companies, and is mostly categorized into four types of analytics: descriptive (describing/summarizing from existing data), diagnostic (determining based on past performance), predictive (using statistical models and ML techniques to predict potential/possible outcomes) and prescriptive (recommend a path/direction/course of action) (Mehta, 2017).

Interestingly, AI-related start-ups have a huge growth potential in India, which became the "second largest start-up hub of the world" (Dewan, 2019). According to Jiten Jain, Director of Voyager Infosec Private Ltd, in India there are certain start-ups that "have done wonderful work especially in the facial recognition, predictive intelligence... [and] that's where a lot innovations are coming" (*Rajya Sabha TV*, 2018b). The AI-related start-ups in the country saw a growth rate of 266% during 2014 to 2020.The prominent start-ups include Niramai Health Analytix (in healthcare, especially in the detection of breast cancer), Haptik.ai (in customer service using "conversational AI"), Discovery AI (to increase productivity using "conversational AI"), Niki.ai (in providing digital services by using "digital localized agent"), Doxper (in healthcare and digitizing the data related to it), CropIn (providing agri-related farming solution), Bash.ai (in human resources and employee functionalities), Avaamo (in customer service through "conversation AI"), Artivatic.ai (to increase efficiency and performance through automated decision-making), SigTuple (to use AI platform to create visual medical data) (Srivastava, 2020). Currently, scientists, researchers and related institutions in India are trying to use AI/ML to model, predict and to help mitigate COVID-related problems in the country (*DNA Web Team*, 2021; Karthikeyan et al., 2021, 1–13; PTI, 2021a). The patent landscape of top ten companies and corporation with AI-based patents from India during 2000 to 2019 are given Fig. 3.2.

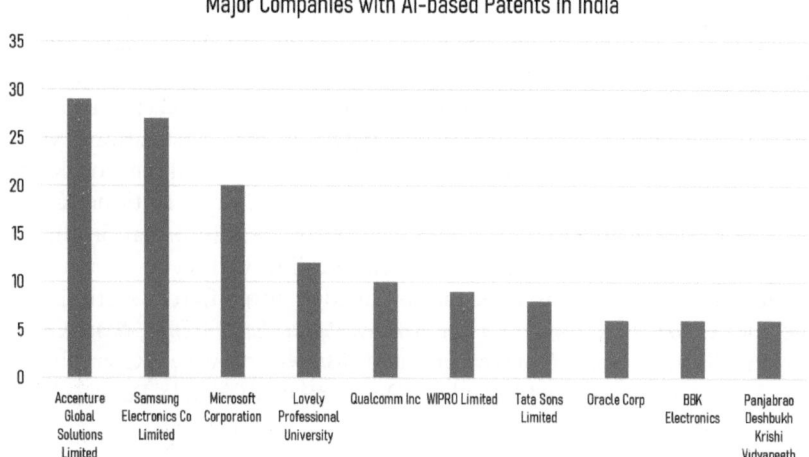

Fig. 3.2 Top ten companies and corporations with AI-based patents from India (2000–19) (*Source* Compiled by Author. *Note* Calculated based on 'ultimate parents')

In terms of State-wise coordination and integration of structures, various states have similar institutional and organizational structures and hierarchical systems in dealing with cyber threats. For the study, the State of Kerala has been taken as a case study here to provide an understanding of the organizational and institutional structures of dealing with cyber threats. The interactions included those with government officials, officers and security professionals.

CYBER POLICING IN INDIAN STATES: A CASE STUDY IN THE STATE OF KERALA

Cyber policing is an important aspect of dealing with cyber threats, especially across states in India, and has become part of initiatives across various states in India, especially the State of Kerala. The State of Kerala became the first state to have a university focusing on cyber technology and digital sciences. The "Kerala University of Digital Sciences, Innovation and Technology (KUDSIT)"[42] at Technocity, Mangalapuram was established in 2000, by the "Indian Institute of Information Technology and Management Kerala (IIITMK)" (*ET Government*, 2021; *The Hindu*, 2021). The overview and analysis are based on field, institutional visits, interactions and websites of the Kerala Police (Department of Kerala Police, 2021), Ministry of Home Affairs (MHA, 2021), National Crime Record Bureau (NCRB, 2021), and archives of Press Information Bureau (PIB, 2021). The governmental agencies dealing with cyber threats and vulnerabilities in the State of Kerala are divided into different sections that include: *General Executive, Crime Branch, Specialized Wings, State Crime Records Bureau, Other Departments*, and *Systems & Projects* (Compiled by the Author based on Field Studies; Department of Kerala Police, 2021; NCRB, 2021).

General Executive

General Executive is further divided into State Police Head Quarters, the various "Police Zones", "Police Ranges" and "Police Districts". Firstly, the State Police Head Quarters is located in Thiruvananthapuram, the State capital, and "is headed by the Director-General of Police (DGP), the State Police Chief". The DGP is assisted in this regard by the "Staff Officers", who help in discharging various duties. Besides, there are some Special Cells, namely the NRI Cell, the Special Cell, the Hi-Tech Crime Enquiry Cell, as well as the Command Centre (Department of Kerala Police, 2021; NCRB, 2021). Secondly, the Police Zones are controlled by the DGP, the highest-ranking officer in the State Police for maintaining law and order. However, "for effective administration, the state is divided into two zones": the "North Zone" and the "South Zone", each of which is headed by an Additional Director General of Police (ADGP). "The office of ADGP (North Zone) is located at Nadakkavu (Kozhikkode District), and the office of ADGP (South Zone) is located at Nandavanam" (Thiruvananthapuram District) (Ibid.).

Thirdly, the Police Ranges are within each of the Police Zone, and is further divided into two Ranges each. Each of the Range is headed by

---

[42] Four key themes are *Computing, AI, Sustainability* and *Entrepreneurship.* The specialization "in digital technologies, including AI, cybersecurity, data analytics, automation, ecological informatics, geospatial analytics, digital economic".

an Inspector General of Police (IGP). Under the North Zone, there are two ranges: Thrissur Range (High Road) & Kannur Range (Thavakkara); and under the South Zone, there are two ranges: Thiruvananthapuram Range (Nandavanam) & Ernakulam Range (at Shanmugham Road). Fourthly, the Police Districts are subdivisions of the Police Ranges, and are numbered (as of December 2018) into 19. For administrative purposes, these have been categorized into five Commissionerates and 14 Police Districts units. The Thiruvananthapuram city and the Kochi city are being headed by the Deputy IGP (DIGP), while the remaining Police Districts are led by an officer of the rank of Superintendent of Police (SP) (Ibid.).

The Police Districts in respective Thiruvananthapuram Range include "Thiruvananthapuram City, Thiruvananthapuram Rural, Kollam City, Kollam Rural, Pathanamthitta"; the Ernakulam Range includes "Alappuzha, Kottayam, Idukki, Ernakulam Rural, Kochi City"; the Thrissur Range includes "Thrissur City, Thrissur Rural, Palakkad, Malappuram"; and finally, the Kannur Range includes "Kozhikode City, Kozhikode Rural, Wayanad, Kannur, Kasaragod". Finally, the City Commissionerates are assisted by officers of the rank of the SP. They are assisted by Assistant SP (ASP) in the city or Deputy SP (DySP) in other regions, "commonly designated as Sub Divisional Police Officers". The Sub Divisional Police are each in charge of a Sub Division. A Sub Division is subdivided into Circles, where Circle Inspectors (CI) are in charge. Each Circle is further divided "into Police Station areas, where a Sub-Inspector with Head Constables and Constables, and if necessary, an additional Sub Inspector to assist" (Ibid.).

Crime Branch

It is further divided into Crime Branch Head Quarters, "Hurt & Homicide Wing", Organized Crime Wing, "Economic Offences Wing, Cyber Crime Police Station, Hi-Tech Crime Enquiry Cell, Anti-Piracy Cell, Special Temple Anti-Theft Squad [and] Internal Security Investigation Team". Firstly, the Crime Branch Head Quarters consists of a specialized investigation wing, known as the Crime Branch Crime Investigation Department (CBCID). It investigates cases "entrusted to it by the State Police Chief, the Government, and/or the High Court of Kerala" (Ibid.).

"It is the nodal agency for Interpol" in relation to "matters in the State and conducts verifications" and/or "enquiries on behalf" of the Interpol.

It is "headed by an officer of the rank" of ADGP. It mainly consists of 3 major wings that are categorized on the basis of cases, and are as follows: "Hurt & Homicide Wing (HHW)", Organized Crime Wing (OCW) and "Economic Offences Wing (EOW)". The HHW had been created in the year 2009. It is led by an officer of an IGP's rank, and consists of units at "Thiruvananthapuram, Ernakulam and Kozhikode". The OCW investigates crimes "committed by a highly organized group of individuals that may be local or spread across various states or nations". It is "headed by an officer of the rank" of an IGP, and is assisted by SPs at the Police Districts of Thiruvananthapuram, Ernakulam, Thrissur and Kannur (Ibid.).

The EOW has been created in 2009, and "investigates various types of frauds that result in wrongful gain to the offender(s) and wrongful loss to the victim(s)". It investigates banking frauds, including credit card and/or debit card frauds, identity theft, money laundering and the like. It is headed by IGP, and consists of units headed by SPs at Thiruvananthapuram, Ernakulam and Kozhikode. Secondly, the Cyber Police Station was established in 2009, to investigate cybercrimes and other related threats permeating through cyberspace. It is headed by "an officer of the rank of DySP and is [being] supervised by SP (Crimes) under the overall command of ADGP (Crimes)". It is empowered to investigate issues like source code theft, cyber terrorism, website hacking, bank account hacking, child pornography, tampering documents related to computer source, social media abuse, related cyber offences through smartphones, "empowered to register FIRs and [even] lay final reports before the Court" (Ibid.).

Thirdly, the Hi-Tech Crime Enquiry Cell (HTCE) began as "a special cell of Kerala Police and started functioning [in] 2006". It has "been created to prevent and detect serious and organized" cybercrimes and other related threats with the assistance of other government agencies. Additionally, collaboration and cooperation with the "private sector, academic institutions, and foreign counterparts" are also undertaken. It "currently functions under the direct supervision of SP (Crimes) and the overall command of" ADGP (Crimes). It is the "nodal unit for the entire Kerala Police to interact with [such] units like" Centre for Development of Advanced Computing (CDAC), "Centre for Development of Imaging Technology (C-DIT), National Informatics Centre (NIC), Kerala [IT Mission], Kerala State Electronics Development Corporation Limited (KELTRON)", Kerala State Audio-Visual & Reprographic

Centre (KSAVRC), Bharat Sanchar Nigam Limited (BSNL), and ISPs. The HTCE enquires into issues like hacking, deformation and disruption of websites and emails, fraudulent online activities, child pornography, financial scamming, phishing activities, digital thefts (of source code and identity), cyber-bullying (including social media abuse) and other digital crimes (that includes the misuse of computer, internet and mobile phone). It provides support, expertise and analytical skills to various police forces to investigate cybercrimes and other threats where related technology has been used. It also undertakes extensive awareness activities on cyber-crimes and other threats, and "its prevention among students, employees of various organizations and the general public" (Ibid.).

Fourthly, "the Anti-Piracy Cell is a specialized wing that investigates copyright infringements, and is headed by an SP. It operates under the overall supervision of ADGP (Crimes)". The Anti-Piracy Cell treats the "source code of a computer programme as a literary work, with certain additional exceptions". The Anti-Piracy Cell works "as a central unit to coordinate" the collection of intelligence, "creating a database, coordinate investigation on copyright offences by local police, and directly investigate important crimes to trace out their (original) source and web distri-bution". Fifthly, the Internal Security Investigation Team was initially created by the name Internal Security Investigation Branch (ISIB). It was later renamed as Internal Security Investigation Team (ISIT) for the exclusive investigation of cases affecting internal security. It is "headed by [an] SP", and functions "under the overall supervision of ADGP (Crimes)". It primarily investigates cases affecting the security of the state, under the Unlawful Activities Prevention Act, by various organizations such as Maoist, communal, fundamentalist groups and/or organization, and even offences that have inter-state or national implications (Ibid.).

Specialized Wings

It is further divided into Cyber Forensic Division and the Crime Wing (under the State Women Cell). Firstly, the Cyber Forensic Division was established to provide for the extraction of information as well as data from a computer, mobile phone and/or any other storage devices and media. The cyber lab is equipped with tools to analyse digital data and to provide valuable information in resolving and neutralizing threats from cyberspace. Secondly, the Crime Wing is one of the three wings under the State Women Cell, and is headed by an SP. It is functioning at Thiru-vananthapuram. The priority is to deal with the grievances of women, especially the issues of harassment, neglect, desertion, lack of recognition of rights and even family discords. This also includes "District Head [Quarters], District Special Branch, District Crime Records Bureau, District Crime Branch, [and Cyber] Cell" (Ibid.).

State Crime Records Bureau Head Quarters

It was established in 1989 as per the directive of the GoI, and is assigned to collect, coordinate, update, store data and train officers in the computer wings. It consists of five divisions, all of which is under the direct control of a DGP. It is further divided into the Crime Intelligence Bureau, "Information and Communication Technology Wing", Police Telecommunications, "Finger Print Bureau, Police Photographic Bureau". Firstly, the Crime Intelligence Bureau is assigned with maintaining and analyzing the statistics of crime in the State. Secondly, the Information and Communication Technology Wing is the nodal agency for undertaking department computerization, provides software and its maintenance. It is supported by the Statistical Wing and Police Computer Center. Statistical Wing is responsible for data collection, statistical analysis and publication; and is headquartered at Thiruvananthapuram, while the Police Computer Center is responsible for maintaining, expanding and modernizing the existing infrastructure in relation to "computerization of the department and development of analytical software as per the requirements of State Police" (Ibid.).

The "Information and Communication Technology Wing" coordinates with the Ministry of Home Affairs (MHA) (including NCRB), GoI and other counterparts "in implementing systems developed commonly for all states as part of the standardization of the police work along with Crime and Criminal Tracking Network & Systems (CCTNS)" that involves collaborations with Price Waterhouse Coopers Private Limited (PwC) at the National-level and Accenture Services Private Limited as the "State Project Monitoring Unit (SPMU)" and Tata Consultancy Services Limited (TCS) as the "System Integrator (SI)" at the State-level. Further, it involves coordination activities along with Common Integrated Police Application (CIPA), Centre for Cyber and Information Security (CCIS), Motor Vehicle Coordination System (MVCS) or Vahan Samanvay, Colour Portrait Building System (PBS), District Crime

Records Bureau (DCRB), Fake Indian Currency Notes (FICN) Information System, Talash Information System and Digital Police Portal, Citizen Portal and the Interoperable Criminal Justice System (ICJS) (Department of Kerala Police, 2021; NCRB, 2021; PIB, 2021).

Thirdly, the Police Telecommunications was formed in 1994 as the successor of the Radio Unit of "Kerala Police is responsible for the maintenance of wireless network of the department".[43] It is considered the backbone of the Kerala Police, and is of utmost strategic significance. It maintains a computer network that connects all the major establishments of the Kerala Police. Fourthly, the State Finger Print Bureau was instituted by the Maharaja Sree Moolam Thirunal Rama Varma of the Travancore Kingdom on May 30, 1900, and is considered one of the oldest in the world.[44] It is located at Thiruvananthapuram, and "19 (Single Digit) Finger Print Bureaux" are located at both district and city levels. It is a "scientific investigation wing" under the Kerala Police, and helps investigative agencies to deal with threats. It also works closely with the "Central Bureau of Investigation (CBI), National Intelligence Agency (NIA)", Interpol and other State Police Forces. It is "headed by a Director and functions directly under" the ADGP ("State Crime Records Bureau"). Finally, the Police Photographic Bureau began in 1935 but started functioning separately only in 1998. The duty and responsibility involve collecting and enhancing digital evidence, including photographs, documents and other data (Field Studies; Department of Kerala Police, 2021; MHA, 2021; PIB, 2021).

Other Departments

The *Other Departments* that deal with cyber threats include the NRI Cell, Police Information Centre and Police Message Centre. Firstly, the NRI Cell functions in Police Headquarters at Thiruvananthapuram under the direct supervision of ADGP (Special Services & Traffic) and SP (NRI Cell). The NRI Cell handles the complaints of NRI's and their family

---

[43] That includes Very High Frequency (VHF) ranging from 30 to 300 MHz, High Frequency (HF) ranging from three to 30 MHz, and Satellite Communication (SATCOM) ranging from one to 300 GHz. SATCOM encompasses the spectrum of Super High Frequency (SHF), Extra High Frequency (EHF) and a small portion of the Ultra High Frequency (UHF).

[44] The world's oldest fingerprint bureau was at Calcutta (modern-day Kolkata) in 1897.

members exclusively, and has jurisdiction over the entire state. Secondly, the Police Information Centre began in 2009 and functions under a Deputy Director of Public Relations, usually on deputation. It works as a media cell as well as an information centre through which various kinds of information are disseminated to the public and the press. Thirdly, the Police Message Centre was launched for the general public to communicate to the police through SMS. "It is intended to prevent crimes, terrorist acts rendering immediate assistance on receipt of the information" on various matters, including harassment by anti-social elements. "An All-Kerala Police Message Centre is functioning" at Police Head Quarters, enabling increased coordination and communication between the public, concerned Police Station(s), CI's Office and DySP's Office, and are put up to the DGP (Department of Kerala Police, 2021; NCRB, 2021).

Systems & Projects

There are other centres of "technological research and development", particularly in collaboration between the Centre of Kerala Police Department and the civilians. The *Kerala Police Cyber Dome* is one such enterprise in enabling and advancing cybersecurity capabilities, and "technological augmentation". This is undertaken through collaboration and partnership between the organizational/institutional actors and a multitude of stakeholders to prevent cyber attacks. Further, it can (purportedly) neutralize cyber threats and synergize inter-departmental cooperation, coordination and interoperability. The stakeholders, besides the government officials, include "academia, research groups, non-profitable organizations, individual experts from community, ethical hackers, private organizations and other law enforcement agencies in the country". It is under the direct supervision of ADGP HQ (Cyber Dome), who is the nodal officer of the *Cyber Dome*, and reports to the DGP. The DGP is the Kerala Convenor of *Cyber Dome* (Kerala Police Cyberdome, 2020).

Cyber Diplomacy and International Cooperation

Cyber diplomacy is an important aspect of international cooperation in cyberspace, and is an important aspect of coherent integration and other related aspects of cyber technological paradigms in India. The key issues include sovereignty, legal jurisprudence, data retention, data preservation, data accessibility, data residency, "formal and informal international cooperation mechanisms, extraterritorial actors" and evidences, national capability and capacity etc. There are some international as well as inter-governmental laws, mechanisms and instruments having international implications in dealing with cyber technological paradigms, which enables cooperation between India and national governments. They include the Electronic Communications Privacy Act (1986), "Computer Fraud and Abuse Act (1986)" of the US, the Privacy Act (1988) of Australia, the Computer Misuse Act (1990) of the UK, Privacy Act (1993) of New Zealand, the "Communications Assistance for Law Enforcement Act or CALEA" (1994) of the US (Reghunadhan, 2018; Sakurayuki et al., 2019; UNODC, 2013; UNODC, 2019; UNODC, 2020).

At the beginning of the twenty-first century, there were further move-ments in terms of international as well as inter-governmental laws, mechanisms and instruments the *"Council of Europe's Convention of Cybercrime* (2001), *Agreement on Cooperation in Combating Offences related to Computer Information"* (2001) *of the Commonwealth of Inde-pendent States, Cybercrime Act* (2001) of Australia, *APEC Privacy Frame-work* (2005), *"Arab Convention on Combating Information Technology Offences"* (2010), *"Agreement on Cooperation in the Field of International Information Security* (2010)" *of the "Shanghai Cooperation Organiza-tion (SCO), Global Programme for Strengthening Capacities of Member States to Prevent & Combat Organized and Serious Crime"* or *GPTOC*[45] (2010 & 2012), *Directive on Fighting Cybercrime*[46] (2011) *of Economic Community of West* African States (ECOWAS), *Model Law on Computer Crime and Cybercrime*[47] (2012) of the Southern African Development Community (SADC), *African Union Draft Convention on the Establish-ment of a Legal Framework Conductive to Cybersecurity in Africa* (Draft African Union Convention) (2012), *Data Privacy Act* (2012) of Philip-pines, the *UNODC Draft Comprehensive Study on Cybercrime* (2013), the

---

[45] It supports and facilitates international networks like "the West African Network of Central Authorities and Prosecutors or WACAP, the Judicial Cooperation Network for Central Asia and Southern Caucasus or CASC, the Great Lakes Regional Judicial Cooperation Network or GLCJN, the South East Asia Justice Network or SEAJust" (UNODC, 2020).

[46] It provides guidelines to *"Convention on Mutual Assistance in Criminal Matters* and the *Convention on Extradition"* (UNODC, 2019).

[47] Serves as guidelines for the implementation of *"SADC Protocol on Mutual Legal Assistance in Criminal Matters* and the *SADC Protocol on Extradition"* (UNODC, 2019).

"Guidelines on the Protection of Privacy and Transborder Flows of Personal Data (2013) *of the Organization for Economic Co-operation and Development (OECD)"*, the "African Union *Convention on Cyber Security and Personal Data Protection"* (2014), *the Data Protection Law* (2016) *of Qatar, Electronic Information and Transactions or EIT Law* (2016) *of Indonesia, the General Data Protection Regulation or GDPR (2018) of the EU,* "the *Convention for the Protection of Individuals with regard to Automatic Processing of Personal Data"* (2018), and *Management of Electronic Systems and Transactions* (2019) of Indonesia. The above laws and mechanisms seek "to harmonize national laws, improve... investigation techniques, and improve international cooperation" (APEC, 2005; UNODC, 2013; UNODC, 2019; UNODC, 2020).

Many countries have increasingly supported the Digital India Programme. India has cooperation with agencies like the "European Union Agency for Law Enforcement Cooperation (EUROPOL), International Criminal Police Organization (INTERPOL)", North Atlantic Treaty Organization (NATO) or even bilaterally as well as multilaterally with different countries take place in the form of agreements and mutual legal assistance treaties (MLATs). Internationally, India has signed the mutual legal assistance treaties (MLATs)[48] with 40 countries including:

Australia (2011), Azerbaijan (2013), Bahrain (2005), Bangladesh (2011), Belarus (2006), Bosnia & Herzegovina (2010), Bulgaria (2008), Canada (1998), Egypt (2009), France (2005), Hong Kong (2009), Iran (2010), Indonesia (2011), Israel (2015), Kazakhstan (2000), Kyrgyz Republic (2014), Kuwait (2007), Malaysia (2012), Mauritius (2006), Mexico (2009), Mongolia (2004), Myanmar (2010), Russia (2000), Singapore (2005), South Africa (2005), South Korea (2005), Spain (2007), Sri Lanka (2010), Switzerland (1989), Sultanate of Oman (2015), Tajikistan (2003), Thailand (2004), Turkey (1993), Ukraine (2003), United Arab Emirates (2000), United Kingdom (1995), United States of America (2005), Uzbekistan (2001), Vietnam (2008), and Maldives (2019) (CBI Systems Division, 2021).

In December 2020, India hosted the 6th India-EU Cyber Dialogue, with Indian and EU contingents led by Joint Secretary, Cyber Diplomacy

---

[48] MLAT is "an agreement between countries to cooperate on investigations and prosecutions of certain and/or all offences proscribed by both parties under national law" (Maras, 2020).

Division (MEA) and Joanneke Balfoort, Director Security and Defence (European External Action Service), respectively. India-EU cooperation in cyberspace, and stability by engagement through "United Nations Group of Governmental Experts (UNGGE)",[49] "Open-Ended Working Group (OEWG)",[50] "Organization for Security and Co-operation in Europe (OSCE)",[51] ASEAN Regional Forum (ARF),[52] EU Cyber Diplomacy Toolbox (CDT),[53] and Cooperation on Cybercrime and Capacity. Besides, the focus of the exchanges between India and the EU included: "cyber policies; Internet Governance; New Emerging cyber-related technologies" (UNODA, 2019; Deka, 2018; Attatfa et al., 2020; MEA, 2020a). The "Framework for the US-India Cyber Relationship" released in June 2016 focused on a particular set of "shared principles". This emphasized state behaviour in cyberspace that entails promotion of

the Internet as an engine for innovation, economic growth, and trade and commerce... cooperation between and among the private sector and government authorities on cybercrime and cybersecurity... bilateral and international cooperation for combating cyber threats and promoting cybersecurity... cultural and linguistic diversity... international security and stability in cyberspace... applicability of international law, in particular the UN Charter... multistakeholder model of internet governance that is transparent and accountable... recognition of the leading role for governments

[49] Since 2004, six working groups have been established. The prominent achievements include "outlining the global agenda" and the application of international law in digital space (*GIP Digital Watch*, 2019).

[50] It was established by the UN General Assembly through resolution 73/27 with Ambassador Jürg Lauber (Switzerland) as Chair (UN, 2021).

[51] It consists of 57 members from North America, Europe and Asia. It is based in Europe, and address issues and security threats to EU, especially "including arms control, terrorism, good governance, energy security, human trafficking, democratization, media freedom and national minorities" (OSCE, 2021).

[52] In 2012, a 19th ARF Ministerial Meeting adopted and developed a workplan by *Statement on Cooperation in Ensuring Cyber Security by the ARF Foreign Ministers* (ARF, 2018). It put forward a workplan to "improve cooperation to respond to criminal and terrorist use of ICT's and to cooperate to develop resilient government ICT environments" (EU Cyber Direct Strategy, 2021).

[53] It includes "*statements & demarches, statements by HR/VP, capacity building, international agreements, strategic communication, joint investigations, EU demarches, a formal request for assistance, Council conclusions, political and cyber dialogues, recalling diplomats, recalling diplomats, solidarity clause, sanctions, countermeasures, Mutual Defence Clause,* and *military response*" (Moret & Pawlak, 2017).

in cybersecurity matters relating to national security... capacity building in cyber security and cyber security research and development... combat cybercrime between the two... protect and respect human rights and fundamental freedoms online... strengthening the security and resilience of critical information infrastructure... skill development... [promotion of norms]... dialogue and engagement in Internet governance fora, including ICANN, IGF and other venues... strengthening critical internet infrastructure in India... ensure shared understanding of technology access policy, including dual use technologies sought to be controlled by either country, including through such mechanisms as the bilateral High Technology Cooperation Group. (MEA, 2016)

In January 2021, the Indian National Security Advisor (NSA)'s interaction with the US emphasized cooperation in combating cybersecurity, with a focus on building "shared values and common strategic and security interests" through expanding the "Common Global Strategic Partnership" (MEA, 2021c). Vietnam was reportedly looking forward to "seek cooperation" in emulating the Digital India Programme. This focused on fields like "cyber security and e-governance", which will have an impact on a quarter of Vietnam's GDP (Ghoshal, 2017). An increasing number of multinational companies (MNCs) have also initiated support and enable institutionalization of the Digital India Programme. Mark Zuckerberg, the Facebook CEO, announced his support for the Digital India Programme. Meanwhile, other global companies like Apple, Google, Microsoft, Qualcomm have all announced their support for the programme (TNN, 2015; Thomas, 2015). Google provided infrastructural and technical support for Wi-Fi connectivity in Indian railways across the country, connecting billions of people at the same time. It was later transferred to the State-owned Rail Tel, which currently connects 5600 railway stations across the country, of which more than seven percent of the railway stations are still powered by the infrastructure of Google (Khan, 2020).

Meanwhile, Microsoft has collaborated with the GoI in enabling "low-cost broadband" connectivity in the rural areas as well as provide support in developing the country with "cloud services through data centres" within. Air Jaldi and Zaya Labs, both Indian companies that provide internet service and solutions, are also collaborating with Microsoft and its Airband Initiative to provide "affordable network". This is an important aspect for providing and enhancing data sovereignty and cyber resilience for India, reducing economic burden for high-speed internet connectivity, to increasingly restructure and democratize governance in the country, improve the quality of education, improve mean schooling years through digital platforms, access to other sectors like healthcare, etc., and more importantly to reduce loopholes and vulnerabilities related to (cyber and data) security and vulnerability (McCord, 2021; Microsoft, 2021; Thomas, 2015).

Since 2016, the US-based Qualcomm, a leader in wireless technology and chip-making, has been conducting the Qualcomm Design in India Challenge to find the best talents in the country. It focuses on supporting start-ups and to increase (potential) innovation capabilities and capacity building of sectors like education,[54] public safety, fishing (or marine livelihood),[55] environmental sustainability,[56] and corporate social responsibility,[57] etc., in the country. The main focus is on transforming "India into a digitally empowered society... [with] support to [develop] India's technology ecosystem". The domains of expertise include the "Smart Infrastructure, Biometric Devices, Medical Technology, Rural IoT" and 5G technology (Qualcomm, 2021). Qualcomm has been providing fiscal,[58] non-fiscal[59] and special incentives[60] as well (Ministry of Commerce & Industry, 2021). Another American electric car giant Tesla has set up a car-manufacturing unit in the country, an important step towards electric and autonomous vehicles and the integration of high-tech R&D on renewables (Kapoor, 2021). US-based Oracle, a leader in "grid computing and data warehousing", has been investing in India, covering 20 states, providing support in cloud services, digital payment

---

[54] Play 'n' Learn app.

[55] Fisher Friend app.

[56] SootSwap Project.

[57] Qualcomm Aqriti Project.

[58] Fiscal incentives include an initial grant (1.6 lakh INR), an additional grant (1.6 lakh INR) and awards to one winner (65 lakh INR), second prize (50 lakh INR) and third prize (35 lakh INR).

[59] Non-fiscal incentives include access to the state-of-the-art Qualcomm Innovation Lab, free incubation support from NASSCOM Centre of Excellence (CoE), accelerator services support to start-ups, financial and technical support from the Indian Government, and support to participate in summits.

[60] Patent Filing Incentive (1.6 lakh INR upto two patent applications), Innovation Commercialization Fund (ICF) of 60 lakh INR with certification and other related support as well.

services and solutions[61] in the country. It includes support to activities related to "Aadhaar Authentication and eKYC, eSign and DigiLocker" (Mohapatra, 2021).

CONCLUSION

The recent globalized multilateralism and fragmentation of cyberspace have also created possibilities for decentralization and democratization of engagements. In the path to becoming a knowledge economy to provide greater accessibility to information (or knowledge) and clarity of processes, India requires a broader platform to support and complement the institutionalization of the Fourth Industrial Revolution. This has been institutionalized through the emergence of the Digital India Programme. In India, this primarily depends on the proportion of the number of internet connectivity, which is a big step to achieve the goals of interconnectivity, digital governance, digital economy, cyber sovereignty, data residency, data localization, etc. The Digital India Programme can transform the current digital economy and the related business environment of the nation. This transformation can be utilized through the use of digital channels as well as social media that can leverage the promotion, collection, dissemination and usage of government services.

The Digital India Programme is such an initiative of the GoI. This includes increasing coordination and interoperability, providing options for skilling and infrastructural development, as well as in undertaking proactive steps to deal with potential threats and vulnerabilities. An important challenge for the emergence of Digital India is the challenge to deal with a large number of ministries, agencies, institutions and organizations. The level of coordination in cyber policing across States is a major challenge, one which needs to be integrated and upgraded as well. The level of upgradation and integration is less evident in state-level institutions, but creates problems for agencies at the Centre. This is exacerbated with the increase in the number of cyber attacks is purportedly originating from its territory, is transforming the balance of power (besides the impact on the relations between different nations) and straining international relations as well. An important directive that the GoI can utilize is the cyber diplomacy and international cooperation-related activities, that

---

[61] It includes the Oracle Startup Cloud Accelerator (OSCA) program, the next-generation Oracle Cloud Infrastructure (OCI), Oracle Dedicated Region, etc.

can spur Indian activities in cyberspace through a cohesive, coordinated and cooperative international cybersecurity architecture.

REFERENCES

ASEAN Regional Forum (ARF). (2018). *ASEAN regional forum work plan on security of and in the use of Information and Communications Technologies (ICTs)*. Retrieved December 10, 2019, from https://aseanregionalforum. asean.org/wp-content/uploads/2018/07/ARF-Work-Plan-on-Security-of-and-in-the-Use-of-Information-and-Communications-Technologies.pdf

Asia-Pacific Economic Cooperation (APEC). (2005). *APEC Privacy Framework*. Retrieved November 21, 2018, from https://www.apec.org/Publications/2005/12/APEC-Privacy-Framework

Attatfa, A., Renaud, K., Paoli, S. D. (2020). Cyber diplomacy: A systematic literature review. 24th International Conference on Knowledge-Based and Intelligent Information & Engineering Systems. *Procedia Computer Science, 176,* 60–69.

Central Bureau of Investigation Systems Division (CBI Systems Division). (2021). *MLATs: MLAT signed by India with other foreign countries as per list provided by MHA*. Retrieved February 16, 2021, from https://cbi.gov.in/MLATs

Centre for Railway Information Systems (CRIS). (2021). *About us*. Retrieved March 26, 2021, from https://cris.org.in/crisweb/design1/index.jsp#home

Christopher, N. (2018). India's first AI research institute opened in Mumbai. *The Economic Times*. Retrieved May 21, 2019, from https://economictimes. indiatimes.com/tech/ites/indias-first-ai-research-institute-opened-in-mumbai/articleshow/63000704.cms

Data Security Council of India (DSCI). (2020). *National cyber security strategy 2020*. Retrieved March 2, 2021, from https://www.dsci.in/sites/default/files/documents/resource_centre/National%20Cyber%20Security%20Strategy%202020%20DSCI%20submission.pdf

Defence Research and Development Organisation (DRDO). (2020). Retrieved August 1, 2020, from https://www.drdo.gov.in/drdo/English/index.jsp?pg=homebody.jsp

Deka, C. (2018). Global Conference on Cyber Space (GCCS) (2017). *Indian Council of World Affairs (ICWA)*. Retrieved February 2, 2020. https://www.icwa.in/show_content.php?lang=1&level=3&ls_id=4916&lid=1738

Department for Promotion of Industry and Internal Trade (DIPP). (2021). *What's new. Ministry of Commerce and Industry*. Retrieved January 29, 2021, from https://dipp.gov.in/

Department of Commerce. (2021). *Annual report 2020–21. Ministry of commerce and industry.* Retrieved March 4, 2021, from https://commerce.gov.in/wp-content/uploads/2021/03/Commerce-English-2020-21.pdf

Department of Electronics and Information Technology (DeitY). (2012). *eDistrict manager fellowship Program: Hiring guidelines.* Retrieved March 1, 2018, from https://www.meity.gov.in/writereaddata/files/e-District_District%20Project%20Manager_Hiring_Guidelines.pdf

Department of Empowerment of Persons with Disabilities (DEPWD). (2021). Retrieved January 29, 2021, from http://disabilityaffairs.gov.in/content/

Department of Kerala Police. (2021). *Home.* Government of Kerala. Retrieved February 21, 2021, from https://keralapolice.gov.in/

Department of Science and Technology (DoST). (2021). Retrieved March 15, 2021, from https://dst.gov.in/

Department of Telecommunications (DoT). (2018). *National Digital Communications Policy—2018.* Retrieved June 11, 2019, from https://dot.gov.in/sites/default/files/Final%20NDCP-2018_0.pdf

Department of Telecommunications (DoT). (2021). Retrieved February 1, 2021. https://dot.gov.in/

Dewan, N. (2019). In the race for AI supremacy, has India missed the bus?. *The Economic Times.* Retrieved July 21, 2020, from https://economictimes.indiatimes.com/small-biz/startups/features/in-the-race-for-ai-supremacy-has-india-missed-the-bus/articleshow/69836362.cms?from=mdr

Dharna. (2020). India now has over 500M active internet users: Google India's Country Head. *Inshorts. Inshorts Medialabs Pvt Ltd.* Retrieved January 10, 2021, from https://inshorts.com/en/news/india-now-has-over-500m-active-internet-users-google-indias-country-head-1594633397013

Dilipraj, E., & Reghunadhan, R. (2017). Organisational framework of India's cyber defence and response. *CAPS-INFocus, 47*(17), 1–5.

Dilipraj, E., & Reghunadhan, R. (2018). Organisational governance of cyberspace in India. *Journal of Air Power and Space Studies, 13*(1), 115–134.

DNA Web Team. (2021). Third COVID-19 wave to strike India by October, IIT Kanpur study predicts. *DNA India.* Retrieved May 12, 2021, from https://www.dnaindia.com/india/report-third-covid19-wave-to-strike-india-by-october-iit-kanpur-study-predicts-2889470

Ecommerce Guide. (2021). *Top 10 ecommerce sites in India 2020.* Retrieved March 1, 2021, from https://ecommerceguide.com/top/top-10-ecommerce-sites-in-india/

Ecourts Services. (2016). *High courts of India, District and Taluka Courts of India.* Retrieved June 12, 2018, from https://ecourts.gov.in/ecourts_home/

ET Bureau. (2021). *TRAI's blockchain solution to curb pesky messages a template for government: Experts.* ET Government. Retrieved March 5, 2021. https://government.economictimes.indiatimes.com/news/digital-india/trais-blockc

hain-solution-to-curb-pesky-messages-a-template-for-government-experts/
81324391

ET Government. (2020). *Labour ministry launches online services via 'Santusht'
portal.* Retrieved June 21, 2020, from https://government.economictimes.
indiatimes.com/news/digital-india/labour-ministry-launches-online-services-
via-santusht-portal/73927155

ET Government. (2021). *India's only digital varsity in Kerala to roll out tech
programmes in online, offline mode.* Retrieved March 4, 2021. https://
government.economictimes.indiatimes.com/news/digital-india/indias-only-
digital-varsity-in-kerala-to-roll-out-tech-programmes-in-online-offline-mode/
81313866

ET Online. (2017). India is quietly preparing a cyber warfare unit to fight a
new kind of enemy. *The Economic Times.* Retrieved June 12, 2018. https://
economictimes.indiatimes.com/news/defence/india-is-quietly-preparing-a-
cyber-warfare-unit-to-fight-a-new-kind-of-enemy/articleshow/61141277.cms

EU Cyber Direct Strategy. (2021). https://eucyberdirect.eu/

Ghoshal, A. (2017). Vietnam may emulate digital India, seeks cooperation on
e-governance. *Hindustan Times.* Retrieved 11 August 2020, from https://
www.hindustantimes.com/india-news

GIP Digital Watch. (2019). UN GGE and OEWG. *Geneva Internet Platform.*
Retrieved February 10, 2020, from https://dig.watch/processes/un-gge

Government of India (GoI). (2020). *AatmaNirbhar Bharat Abhiyan.* Retrieved
January 15, 2021, from https://aatmanirbharbharat.mygov.in/

Hebbar, P. (2018). Top 6 AI collaborations between NITI Aayog and tech giants
In 2018. *AnalyticsIndiaMag.* Retrieved June 1, 2019, from https://analytics
indiamag.com/niti-aayog-ai-collaborations-2018/

India Brand Equity Foundation (IBEF). (2014). Vision for a digitally empowered India. *Ministry of Commerce & Industry.* Retrieved May 10, 2018, from
https://www.ibef.org/blogs/vision-for-a-digitally-empowered-india

India Brand Equity Foundation (IBEF). (2020a). Infrastructure. *Ministry of
Commerce & Industry.* Retrieved December 2, 2020, from https://www.ibef.
org/download/Infrastructure-November-2020.pdf

India Brand Equity Foundation (IBEF). (2020b). MOU signed between NeGD
and CSC E-governance services india limited making services on UMANG
app available to citizens through the network of 3.75 lakh CSCs. *Ministry
of Commerce & Industry.* Retrieved March 2, 2021, from https://www.ibef.
org/news/mou-signed-between-negd-and-csc-egovernance-services-india-
limited-making-services-on-umang-app-available-to-citizens-through-the-net
work-of-375-lakh-cscs

India Brand Equity Foundation (IBEF). (2021a). E-commerce industry in
India. *Ministry of Commerce & Industry.* Retrieved February 12, 2021, from
https://www.ibef.org/industry/ecommerce.aspx

India Brand Equity Foundation (IBEF). (2021b). Manufacturing sector in India. *Ministry of Commerce & Industry*. Retrieved March 1, 2021, from https://www.ibef.org/industry/manufacturing-sector-india.aspx

Indo-Asian News Service (IANS). (2021a). *Cisco, Nasscom to offer 20K virtual internships on cyber security*. ET Government. Retrieved March 3, 2021, from https://government.economictimes.indiatimes.com/news/digital-india/cisco-nasscom-to-offer-20k-virtual-internships-on-cyber-security/81273239

Indo-Asian News Service (IANS). (2021b). *Nasscom partners with Microsoft, announces 'AI Gamechangers' programme*. ET Government. Retrieved March 5, 2021, from https://government.economictimes.indiatimes.com/news/digital-india/nasscom-partners-with-microsoft-announces-ai-gamechangers-programme/81306530

Institute of Chartered Accountants of India. (2021). Retrieved January 19, 2021, from https://www.icai.org/

Institute of Company Secretaries of India (ICSI). (2021). *ICSI-empowering new India booklet*. Retrieved January 19, 2021, from https://www.icsi.edu/media/webmodules/ICSI-Empowering_New_India_Booklet.pdf

Institute of Cost Accountants of India. (2021). Retrieved January 19, 2021, from https://icmai.in/Advanced_Studies/DAFP/index.php

International Monetary Fund (IMF). (2021). *World economic outlook update*. Retrieved February 3, 2021, from https://www.imf.org/en/Publications/WEO/Issues/2021/01/26/2021-world-economic-outlook-update

Kapoor, M. (2021). Tesla to set up India's first manufacturing unit in Karnataka, Says Yediyurappa. *Bloomberg Quint*. Retrieved March 1, 2021, from https://www.bloombergquint.com/business/tesla-to-set-up-indias-first-manufacturing-unit-in-karnataka-says-yediyurappa

Karmanov, G., & Hudson, S. (2018). Global AI talent report (2019). *jfgagne*. Retrieved November 2, 2019, from https://jfgagne.ai/talent-2019/

Karthikeyan, A., Garg, A., Vinod, P., & Priyakumar, U. (2021). Machine learning based clinical decision support system for early COVID-19 mortality prediction. *Frontiers in Public Health, 9*(626697), 1–13.

Keelery, S. (2020a). India social network penetration 2015–2025. *Statista*. Retrieved December 5, 2020, from https://www.statista.com/statistics/240960/share-of-indian-population-using-social-networks/

Keelery, S. (2020b). Internet usage in India—Statistics & facts. *Statista*. Retrieved December 5, 2020, from https://www.statista.com/topics/2157/internet-usage-in-india/?

Kerala Police Cyberdome. (2020). *About Cyberdome*. Retrieved January 10, 2021, from https://cyberdome.kerala.gov.in/about.html

Khan, D. (2020). *RailTel to continue free WiFi service at all railway stations without Google support*. ET Telecom. Retrieved August 21, 2020, from https://telecom.economictimes.indiatimes.com/news/railtel-to-continue-free-wifi-service-at-all-railway-stations-without-google-support/74176766

Kumar, V. (2019). Top artificial intelligence salaries in India in November 2019. *AnalyticsInsight*. Retrieved December 21, 2019, from https://www.analytics insight.net/top-artificial-intelligence-salaries-in-india-in-november-2019/

Kumar, V. (2021a). Top 10 AI companies twitter accounts to follow in 2021. *AnalyticsInsight*. Retrieved May 1, 2021, from https://www.analyticsinsight. net/top-10-ai-companies-twitter-accounts-to-follow-in-2021/

Kumar, V. (2021b). Top artificial intelligence jobs to apply in April 2021. *AnalyticsInsight*. Retrieved May 1, 2020, from https://www.analyticsinsight.net/ top-artificial-intelligence-jobs-to-apply-in-april-2021/

Kumar, V. (2021c). Visualizing tech giants' artificial intelligence ambitions. *AnalyticsInsight*. Retrieved May 1, 2021, from https://www.analyticsinsight. net/visualizing-tech-giants-artificial-intelligence-ambitions/

Langa, M. (2020). Minimum government, maximum governance: PM Modi's mantra to IAS probationers. *The Hindu*. Retrieved December 19, 2020, from https://www.thehindu.com/news/national/minimum-government-maximum-governance-pms-mantra-to-ias-probationers/article32990257.ece

Maras, M. H. (2020). *Cyberlaw and cyberliberties*. Oxford University Press. [Forthcoming].

McCord, M. (2021). This project wants to bring broadband—And hope—To rural India. *World Economic Forum*. Retrieved February 20, 2021, from https://www.weforum.org/agenda/2021/02/digital-divide-rural-india-internet-access-microsoft/

Mehta, A. (2017). Four types of business analytics to know. *AnalyticsInsight*. Retrieved November 2, 2018, from https://www.analyticsinsight.net/four-types-of-business-analytics-to-know/

Menon, S. (2018). Security strategies for India as an emerging regional power with global ambitions. *The United Service Institution of India*. Retrieved December 19, 2019, from https://usiofindia.org/publication/usi-journal/security-strategies-for-india-as-an-emerging-regional-power-with-global-amb itions/

Microsoft. (2021). *Corporate social responsibility*. Retrieved January 21, 2021, from https://www.microsoft.com/en-us/corporate-responsibility/airband

Ministry of Agriculture Cooperation & Farmers Welfare (MoA&FW). (2021). *DACFW organization*. Retrieved February 25, 2021, from http://agricoop. nic.in/hi#skipCont

Ministry of Commerce and Industry. (2021). *Incubator program*. Retrieved April 11, 2021, from https://www.startupindia.gov.in/content/sih/en/ams-applic ation/incubator-program.html?applicationId=5fa38e9ae4b0a5ca1b8e4a8e

Ministry of Education (MoE). (2021). Retrieved January 21, 2021, from https://www.education.gov.in/hi

Ministry of Electronics and Information Technology (MeitY). (2020). *IT minister launches national AI portal of India.* www.ai.gov.in: Inaugurate Intel India's Initiative to Promote AI among School Students. *Press Information Bureau.* Retrieved January 21, 2021, from https://pib.gov.in/PressRelease Page.aspx?PRID=1627950

Ministry of Electronics and Information Technology (MeitY). (2021). *Digital India.* Retrieved January 10, 2021, from https://www.digitalindia.gov.in/

Ministry of External Affairs (MEA). (2016). *Fact sheet on the framework for the US-India cyber relationship.* Retrieved May 1, 2020, from https://mea.gov.in/outoging-visit-detail.htm?26880/Fact+Sheet+on+the+framework+for+the+USIndia+Cyber+Relationship

Ministry of External Affairs (MEA). (2020a). *6th India-EU cyber dialogue.* Retrieved May 11, 2021, from https://www.mea.gov.in/press-releases.htm?dtl/33308/6th_IndiaEU_Cyber_Dialogue

Ministry of External Affairs (MEA). (2020b). *MEA's digital diplomacy footprint.* Retrieved January 11, 2021, from https://mea.gov.in/Portal/IndiaArticle All/636475777798559414_29120_MEAs_Digital_Diplomacy_Footprint.pdf

Ministry of External Affairs (MEA). (2020c). *SEWA—Indian consular services system.* Retrieved December 11, 2020, from https://indianconsularservices.mea.gov.in/consularServices/

Ministry of External Affairs (MEA). (2021a). *About us.* Retrieved March 25, 2021, from https://www.mea.gov.in/divisions.htm

Ministry of External Affairs (MEA). (2021b). *Integration of digilocker platform with passport services.* Retrieved March 2, 2021, from https://mea.gov.in/press-releases.htm?dtl/33552/Integration+of+DigiLocker+platform+with+Passport+Services

Ministry of External Affairs (MEA). (2021c). *Telephonic conversation between national security adviser and Mr. Jake Sullivan, National Security Advisor of U.S.A.* Retrieved February 2, 2021, from https://mea.gov.in/press-releases.htm?dtl/33417/Telephonic+Conversation+between+National+Security+Adviser+and+Mr+Jake+Sullivan+National+Security+Advisor+of+USA

Ministry of Finance (MoF). (2017). *Union budget of India.* Retrieved February 2, 2019, from https://www.indiabudget.gov.in/budget2017-2018/budget.asp

Ministry of Finance (MoF). (2021). Retrieved February 26, 2021, from https://dea.gov.in/

Ministry of Health & Family Welfare (MoHFW). (2021). *Resources.* Retrieved January 1, 2021, from https://www.mohfw.gov.in/

Ministry of Home Affairs (MHA). (2021). *Home.* Retrieved February 11, 2021. https://www.mha.gov.in/

Ministry of Labour & Employment (MoL&E). (2021c). Retrieved February 13, 2021, from https://labour.gov.in/

Ministry of Panchayati Raj (MoPR). (2019). *Year end review-2019: Ministry of Panchayati Raj.* Retrieved May 12, 2020, from https://pib.gov.in/PressRele aseIframePage.aspx?PRID=1598057

Ministry of Panchayati Raj (MoPR). (2021). Retrieved March 31, 2021. https:// egramswaraj.gov.in/

Ministry of Personnel, Public Grievances and Pensions (MoPPGP). (2017). *Estimates and performance review of all India services: Committee on estimates (2016–17) Twenty Sixth Report.* Lok Sabha Secretariat. Retrieved January 21, 2020, from https://eparlib.nic.in/bitstream/123456789/762 126/1/16_Estimates_26.pdf

Ministry of Personnel, Public Grievances and Pensions (MoPPGP). (2020). *Annual report 2019–20.* Retrieved February 21, 2021, from https://dopt.gov.in/sites/default/files/AR%202019-20%20English.pdf

Mohapatra, A. (2021). *'Digital India' gets a push as oracle cloud gets validated for India stack.* Techseen. Retrieved March 25, 2021, from https://techseen.com/oracle-cloud-digital-india/

Moret, E. & P. Pawlak. (2017). The EU cyber diplomacy toolbox: Towards a cyber sanctions regime? *Issue Brief.* Retrieved September 21, 2019, from https://www.iss.europa.eu/sites/default/files/EUISSFiles/Brief%2024%20Cyber%20sanctions.pdf

National Crime Records Bureau (NCRB). (2021). https://ncrb.gov.in/

National e-Governance Division (NeGD). (2021). *Programme management.* Retrieved February 3, 2021, from https://negd.gov.in/node/70

National Institution for Transforming India (NITI Aayog). (2020). *National strategy on artificial intelligence.* Retrieved January 17, 2021, from http://niti.gov.in/index.php/national-strategy-artificial-intelligence#:~:text=Recogn ising%20AI's%20potential%20to%20transform,in%20new%20and%20emer ging%20technologies

National Judicial Data Grid (NJDG). (2020). Retrieved June 12, 2020, from https://njdg.ecourts.gov.in/hcnjdgnew/

National Mission on Interdisciplinary Cyber-Physical Systems (NM-ICPS). (2021). *About—National mission on interdisciplinary cyber physical system.* Retrieved January 13, 2021, from https://nmicps.gov.in/Home/ICPSNM HOME/Aboutus

National Payment Corporation of India (NPCI). (2021). Retrieved January 13, 2021, from https://www.npci.org.in/

National Prisons Information Portal (NPIP). (2021). *Digital innovation in correctional homes.* Retrieved January 13, 2021, from https://eprisons.nic.in/Public/Home.aspx

National Skill Development Corporation (NSDC). (2021). *What we do.* Retrieved March 12, 2021, from https://www.nsdcindia.org/

NSDL Database Management Limited (NDML). (2021). Retrieved March 12, 2021, from https://www.ndml.in/index.php

Organization for Security and Co-operation in Europe (2021). https://www.osce.org/

Paganini, P. (2014). Project NETRA—The Indian internet surveillance. *Security Affairs*. Retrieved June 12, 2018, from https://securityaffairs.co/wordpress/20991/intelligence/project-netra-indian-surveillance.html

Press Information Bureau (PIB). (2014). *Digital India—A programme to transform India into digital empowered society and knowledge economy*. Retrieved January 16, 2020, from https://pib.gov.in/newsite/PrintRelease.aspx?relid=108926

Press Information Bureau (PIB). (2019). *India, a bright spot in global economy*. Retrieved January 14, 2020, from https://pib.gov.in/newsite/PrintRelease.aspx?relid=187451

Press Information Bureau (PIB). (2021). *Archives*. Retrieved February 8, 2021, from https://www.pib.gov.in/indexd.aspx

Press Trust of India (PTI). (2021a). DRDO develops AI tool for COVID detection in chest X-rays. *The Economic Times*. Retrieved May 15, 2021, from https://economictimes.indiatimes.com/news/india/covid-19-vaccination-for-18-44-age-group-suspended-in-mumbai/articleshow/82585713.cms

Press Trust of India (PTI). (2021b). *Nasscom unveils second edition of mentoring programme to promote Indian deep-tech companies*. ET Government. Retrieved March 15, 2021, from https://government.economictimes.indiatimes.com/news/digital-india/nasscom-unveils-second-edition-of-mentoring-programme-to-promote-indian-deep-tech-companies/81342529

Prime Minister's Office (PMO). (2021). *Digital India*. Retrieved February 8, 2021, from https://www.pmindia.gov.in/en/tag/digital-india/#skip_to_main

Programme Management Information System (PMIS). (2021). *Objectives*. Retrieved February 3, 2021, from https://pmis.negd.gov.in/

PRSIndia. (2020). *Summary of announcements: Aatma Nirbhar Bharat Abhiyaan*. Retrieved January 2, 2021, from https://www.prsindia.org/report-summaries/summary-announcements-aatma-nirbhar-bharat-abhiyaan

Qualcomm. (2021). *Empowering the startup ecosystem in India*. Retrieved January 21, 2021, from https://www.qualcomm.com/company/locations/india/design-in-india-program

Rajya Sabha TV. (2018a). In depth—Artificial intelligence in India. *YouTube*. Retrieved November 21, 2019, from https://www.youtube.com/watch?v=CVMjIiry5As

Rajya Sabha TV. (2018b). The big picture—Humans & artificial intelligence. *YouTube*. Retrieved November 21, 2019, from https://www.youtube.com/watch?v=21glMm5RmSI

Rastogi, A. (2020). National Digital Health Mission (NDHM). *National health portal admin.* Retrieved December 21, 2020, from https://www.nhp.gov.in/national-digital-health-mission-(ndhm)_pg#:~:text=The%20Ministry%20of%20Health%20and,health%20infrastructure%20in%20the%20country.&text=The%20NDHM%20is%20a%20collaborative%20initiative%20between%20m any%20ministries%2Fdepartments

Reghunadhan, R. (2018). Cyber threat landscape of digital India: A critical perspective. *Journal of Polity and Society, 10*(1&2), 37–50.

Sakurayuki, C. C. C., Fauziah, I., & Virgiany, M. (2019). *Indonesia's electronic systems and transactions regulation replaced and data regulation amended.* Lexology. Retrieved December 1, 2019, from https://www.lexology.com/library/detail.aspx?g=ab893227-2c51-4090-908d-b3e3cb851761

Sanyal, A. (2020). Bharat surfing more than India, with cheap internet and smartphones. *NDTV.* Retrieved December 25, 2020, from https://www.ndtv.com/india-news/bharat-surfing-more-than-india-with-cheap-internet-and-smartphones-2224508

Saraswat, V. K. (2019). *Cyber security.* NITI Aayog. Retrieved June 14, 2020, from https://niti.gov.in/sites/default/files/2019-07/CyberSecu rityConclaveAtVigyanBhavanDelhi_1.pdf

Sinha, S. (2018). Where artificial intelligence research in India is heading. *Analytics India Magazine.* Retrieved November 20, 2019, from https://ana lyticsindiamag.com/where-artificial-intelligence-research-in-india-is-heading/

The Hindu. (2021). State's digital university opens today. Retrieved February 25, 2021. https://www.thehindu.com/news/national/kerala/states-digital-university-opens-today/article33883848.ece

Srivastava, S. (2019). Top 20 B. Tech in artificial intelligence institutes in India. *AnalyticsInsight.* Retrieved November 29, 2020, from https://www.analytics insight.net/top-20-b-tech-in-artificial-intelligence-institutes-in-india/

Srivastava, S. (2020). Top 10 disruptive AI startups in India 2020. *AnalyticsInsight.* Retrieved December 21, 2020, from https://www.analyticsinsight.net/top-10-disruptive-ai-startups-in-india-2020/

Telecom Disputes Settlement and Appellate Tribunal (TDSAT). (2021). Retrieved February 1, 2021, from https://tdsat.gov.in/Delhi/Delhi.php

Telecom Regulatory Authority of India (TRAI). (2021). *Annual report.* Retrieved March 3, 2020, from https://www.trai.gov.in/sites/default/files/Annaul_Report_02032021.pdf

Thomas, T. K. (2015). Modi effect: Silicon Valley giants commit to digital India. *The Hindu Business Line.* Retrieved February 1, 2018, from https://www.thehindubusinessline.com/info-tech/modi-effect-silicon-valley-giants-commit-to-digital-india/article7694877.ece

Tiezzi, S. (2020). China's bid to write the global rules on data security. *The Diplomat.* Retrieved December 14, 2020, from https://thediplomat.com/2020/09/chinas-bid-to-write-the-global-rules-on-data-security/

Times News Network (TNN). (2015). Mark Zuckerberg changes his profile picture to support 'Digital India'. *The Times of India.* Retrieved December 19, 2019, from https://timesofindia.indiatimes.com/tech-news/Mark-Zuckerberg-changes-his-profile-picture-to-support-Digital-India/articleshow/49128369.cms?pcode=461

United Nations (UN). (2021). *Open-ended working group.* Retrieved February 2, 2021, from https://www.un.org/disarmament/open-ended-working-group/

United Nations for Disarmament Affairs (UNODA). (2019). *Fact Sheet: Developments in the field of information and telecommunications in the context of international security.* Retrieved August 13, 2020, from https://unoda-web.s3.amazonaws.com/wp-content/uploads/2019/07/Information-Security-Fact-Sheet-July-2019.pdf

United Nations Office on Drugs and Crime (UNODC). (2013). *The UNODC global programme for strengthening capacities to prevent and combat organized crime.* Retrieved May 22, 2018, from https://www.unodc.org/documents/organized-crime/GPTOC/13-83720_GPTOC_Approval_nd.pdf

United Nations Office on Drugs and Crime (UNODC). (2019). *The doha declaration: Promoting a culture of lawfulness.* Retrieved January 1, 2020, from https://www.unodc.org/ji/

United Nations Office on Drugs and Crime (UNODC). (2020). *International cooperation in criminal matters.* Retrieved December 21, 2020. https://www.unodc.org/unodc/en/organized-crime/international-cooperation.html#:~:text=Transnational%20organized%20crime%20requires%20a%20coordinated%20transnational%20response.&text=International%20cooperation%20against%20organized%20crime,and%20security%2C%20not%20surrendering%20it

CHAPTER 4

Conclusion: Reimagining India

Abstract The chapter summarizes as well as suggests policy directions to be undertaken in strengthening, improving and elevating the systematization of the Digital India programme.

Keywords Digital India · Policy recommendations · Cyber governance · Reimagining India

INTRODUCTION

Cyberspace has emerged as a mainstream platform that has transformed and revolutionized the underpinnings of human civilization, somewhat similar to the impact of the invention of fire and the wheel. Also referred to as the digital or virtual world, it is part and parcel of what is considered similar to the impact of the Information and Communication Technology (ICT) revolution on the global commons and societies. As a form of social space, cyberspace is considered a public good or resource, which should "normally" be accessible to all. In the creation of cyberspace as a discipline, it has been a hugely popular topic, especially in relation to the interlinkages with a multitude of academic disciplines, for many decades now. In the academic fields, cyberspace research has arisen to provide a wider and broader perspective often that is unforeseen or hitherto heard

© The Author(s), under exclusive license to Springer Nature   111
Singapore Pte Ltd. 2022
R. Reghunadhan, *Cyber Technological Paradigms and Threat Landscape in India*, https://doi.org/10.1007/978-981-16-9128-7_4

of earlier. It has often been argued that the study of fields like defence, strategic and security studies and cyberspace is an undeniable force to reckon with (Sivanesan, 2018; PwC, 2017).

During the aftermath of the 9/11 attacks in the US, Jacques Derrida, the French philosopher, argued that these attacks were still part of the "archaic theater of violence" of the real visible world, and events are still to be conducted in a "clear and great order" (Dilipraj & Reghunadhan, 2018). There is literally no activity that does not involve the use of an electronic device and its connection to the computer networks, thus enabling the connection of cyberspace to almost every nook and corner of the lives of the user. Cyber threats have emerged as one of the important dimensions of threats in (and from) cyberspace, ensuing greater impact on the common netizens more often than the institutional actors. It has become a widespread phenomenon but often is (mis)represented or remains vague due to the issues of conventional conceptualizations, misrepresentations or undue delay, and even loopholes in mechanisms that deal with it. It creates serious problems for modern-day society, encroaching into national security, into netizens' lives, stealing, mining and (mis)using their information leading to psychological and/or physical impact on individuals, organizations, companies and governments.

The huge impact and extent of the perpetuation of malignant and/or malicious online activities can be seen in the likes of Denial of Service (DoS) and the Distributed Denial of Service (DDoS). According to the Federal Bureau of Investigation (FBI), "Cyber intrusions [and attacks] are becoming more commonplace, more dangerous, and more sophisticated... [targeting a] nation's critical infrastructure... companies... for trade secrets". In addition, the use of "sensitive corporate data... [and] universities for... cutting-edge research and development. Citizens are [being] targeted by fraudsters and identity thieves... children are targeted by online predators". It categorizes its key priorities as dealing with computer and network intrusions, ransomware and related priorities like going dark, identity theft and online predators (FBI, 2018).

A cyber threat can (potentially) alter the power structure of international relation(s), transforming overnight the capabilities of a leading power into an underdeveloped nation or even manipulate the democratic institutions and mechanisms of a country, possibly hacking elections and thereby altering the future of not just a nation, but the world as a whole. A country with such a large number of internet users, India needs to have a strong ICT infrastructural base, coupled with the utilization and integration of relatively low wage labour. This provides the country with a huge comparative advantage over developed nations but increases possibilities for further and future international collaborations in dealing with threats. The alternative option(s) include enabling resilience by creating options of data localization, data residency and data securitization to provide for backup and provide for a swift response to the sources of attacks (which is in line with the cyber deterrence strategy in case of cyberwarfare from rival actors). This includes enabling the effective legislative mechanisms, investigative capabilities and responsive authority(ies) and institution(s) in place. In addition, nations could create CNIs as a counter value and/or counterforce against cyber attacks.

Cyber attacks utilize the phone systems globally connected; the world's largest computer network called the Internet. This provides for relatively easy access to reach out to any place at any time as long as there exists accessibility to a communication link(s) and/or phone connection. The spread of satellite and wireless technology has made the importance of location or region irrelevant; where a person exists. Thus, it circumvents the issue of barriers. Phishing attacks are common examples of exploitative attacks on netizens and institutions. It involves targeting a user of the Internet to click on a link through email, webpage or an SMS leading to covertly downloading malicious software to the respective user's device. This helps the attacker gain control over the device and access the information within the device and the network of devices connected to it, both physically and virtually. Incidentally, there is a need to focus on reimagining India's strategies, policies and initiatives in a much more effective and efficient manner, revitalizing and (re)synergizing the country and people.

REIMAGINING INDIA: POLICY RECOMMENDATIONS AND SUGGESTIONS

There are various policy recommendations for the GoI to follow, and they can be categorized as human capital, infrastructure development, technological innovation, smart/digital governance, data/cyber sovereignty.

The first policy suggestion is related to *human capital*, whereby India is among the most populous countries behind only China. Thus, India is home to large personnel with potential, trained and skilled to gain humungous traction in dealing with cyber threats. This is also a priority of the pillars under the Digital India programme, under the *IT for Jobs*. Further, there is a need to update India's cybersecurity strategy. The *National Cyber Security Strategy 2020* brought by DSCI entails capability and skill building among the citizens as an important step forward. Unlike the conventional understanding of working age (say between 15 to 64 years), "digital working age" and "digital literacy", "digital skilling" and specialization in cyberspace or cybersecurity can vary, very much, depending on the foundational understanding and skillsets of an individual or community of individuals. Interestingly, advanced cybersecurity skillsets are easily imbibed during early age, evident in personnel skill training in countries like Israel, the US, UK and other European countries.

The emphasis needs to be on research in cybersecurity is something of greater interest for countries across the world, and India should work as well. Besides, the awareness of cyber hygiene and gender sensitization is important to increase digital best practices among organizations, agencies, private companies and netizens within the country. Further, India should focus on increasing coordination across various ministries, organizations and private agencies to accelerate skilling, recruitment and digital literacy programmes. A vision under the Digital India programme, the *Digital Empowerment of Citizens* emphasizes universality and accessibility for digital literacy and related digital resources (including documents or certificates) through the cloud, availability of Indian languages-based resources and services, enabling participative governance through collaborative digital platforms, and options for portability of all individual entitlements through cloud technology. A digital education policy is necessary to complement the existing National Digital Literacy Mission under the Digital India programme. Moreover, skilling under the Skill India programme, coordination among various ministries and agencies like NASSCOM and AICTE, and collaboration with private companies like Microsoft, Accenture, Google and Facebook has been initiated by the Centre, and State governments need to be expanded to a larger population.

The second policy suggestion is related to the huge issue in *infrastructure development*, viz.*e*.viz., the integration, implementation and deployment of state-of-the-art mechanisms are prerequisites for elevating India as a digital superpower. The Indian digitization process focuses on governance, access through websites and portals, and last-mile connectivity for enhanced accessibility. However, the analyses and inferences of the digitization process, even in economically advanced countries, have had security loopholes and vulnerabilities. Hence, the basic issue of the digital divide should be dealt with, including a focus on expanding *internet infrastructure*. This is a vision under the *Infrastructure as Utility to Every Citizen*, which emphasizes high-speed internet to all Gram Panchayats, providing digital identity, mobile (digital) banking, easier accessibility to local Common Services Centre (CSC), public cloud facilities and ensuring safe and secure cyberspace. Further, digital governance and the digital economy in India unequivocally depend on infrastructure development. The focus on digital governance entailed as *Governance and Services on Demand* emphasizes on seamless integration of various departments or jurisdictions (single window access), real-time availability of government services, accessibility for citizens to the cloud, digital transformation of government services, e-transfer and cashless transactions in the financial sector, and the utilization of the Geographic Information System (GIS) to decision-making and development systems.

This is part of the pillars of the Digital India programme, especially under three pillars: *Broadband Highways, Universal Access to Mobile Connectivity and Public Internet Access Programme*. Telemedicine, e-commerce, tele-education and the social media market in the country are considered to be among the fastest-growing in the world and thus will play a huge role during the (post-)pandemic period. In the age of IoT, AI, blockchain and big data, the government should focus on increasing interoperability, standardization and coordination between NOFN, BSNL, MTNL, and infrastructural components like NKN, SWAN, NOFN, GUN and MeghRaj Cloud. In addition, an important focus should be on improving internet connectivity, and cyber-related defences of websites, internet portals and infrastructure in India need to be strengthened against cyber attacks, with an increase in resilience against attribution-related activities from rival nations. This is particularly important for critical information infrastructure protection (CIIP), including threats to electronic systems like Programmable Logic Controllers (PLC), Remote Terminal Unit (RTU), Supervisory control and data acquisition (SCADA), Distributed Control System (DCS), and Human Machine Interfaces (HMI). The government has been cooperating with foreign and domestic private companies, increasing the quality of infrastructure development in the country.

The third policy suggestion is related to the country's focus on *technological innovation*, which has been seriously considered by the Indian

State as required to provide a huge spur in advancing developmental growth and projecting national power in the South Asian region and the Asia–Pacific region. The focus should be on increasing interlinkages *between academia-industry-government-public environment* as a foundation for elevating India's growth into a "digital knowledge economy". This was described by Carayannis and Campbell (2009) as the "quadruple helix", an extension to the "triple helix model of innovation", an important direction of the "fractal innovation ecosystem" in the twenty-first century (Carayannis & Campbell, 2009). The TIHs in the country need to expand, and strengthen interlinkages with academia, government, civil society and private sector companies. The emergence of frontier technologies like IoT, AI, 3-D printing and big data-related technologies will unequivocally provide a huge impetus to unlocking India to become a major stakeholder in the global innovation hub, with an estimated economic value of 500 billion USD of Indian GDP by 2025 (IANS, 2021).

Further, innovations in the payment and financial systems in the country, with the involvement of ministries, RBI, NPCI and NASSCOM, through various initiatives are expected to enable accessibility, interoperability, cost-effectiveness, integration, effective management of complaints and disputes, standardization and scalability. This can improve the efficacy in effectively transforming the Digital India initiative across various sectors like healthcare, fintech, education, manufacturing, agriculture and other services. In addition, this is linked to enabling companies and corporations to innovate and grow in the domestic and international market, particularly for start-ups. Therefore, an important initiative that the government needs to focus on is providing an ecosystem of improving the Ease of Doing Business in the country while incentivizing domestic companies to engage in R&D (both basic and applied research) and joint ventures with prominent international companies.

A priority list of foreign companies based on the market and IP dominance can provide domestic private sector companies and public sector enterprises (PSEs) in engaging with them. The current focus should be major foreign companies holding AI-based patents in India, in the likes of Accenture, Samsung, Microsoft, Qualcomm and Oracle. The engagement of academic institutions like Lovely Professional University, Panjabrao Deshmukh Vidyapeeth and private domestic companies like WIPRO, Tata in AI-based patenting should be encouraged through (non-)monetary incentives and support. The conducting of hackathons and innovation challenges through foreign private companies as well as private sector agencies in the country should be encouraged. The government should also ensure recruiting talents to the government services as technology and innovation consultants, provide support for start-ups and include investments as an angel or seed investor.

The fourth policy suggestion is related to *smart/digital governance* should focus on digitizing institutions, organizations and information in the country. This is the foundation of the Digital India programme, mainly enabling participative governance through collaborative digital platforms. India can be part of the international institutionalization taking place at various parts of the world, and integrate it at the domestic level, providing a huge platform in transforming the conventional mainstream understanding of governance, governability and access to public goods. Thus, it can help achieve what Prime Minister Narendra Modi believes to be "minimum government and maximum governance" (Langa, 2020). The preparedness and resilience of digital systems are directly linked to governance structures and institutions, of which bureaucracy and judiciary play an important role. The issues and problems of using Aadhaar for linking citizens of the country has only created mainstream narratives and counter-narratives on the "security vs privacy debate", and have been an object of contention for a long period (Reghunadhan, 2018). This debate has only intensified with the recent issue of Whatsapp's privacy policy, one which the Indian State should tread in a very careful and balanced manner before finalizing the issue. The government should, besides the encouragement for competitors in the Indian market, in the likes of Signal or Telegram, commercialize its government-owned app like SANDES (Reghunadhan, 2021).

Implementing bureaucratic and administrative reforms to transform digital governability and smart systems in the country is a prerequisite. This should target administrative logjam, the prevalence of red-tapism and any form of performance stagnation as well. The digitization of bureaucratic and governance-related connectivity will transform the activities of various departments, viz. e. viz., at the panchayat, taluk, district, state and national level. NII is the nodal department with infrastructural components like NKN, SWAN, NOFN, GUN and MeghRaj Cloud (PMO, 2021; Reghunadhan, 2018). A reinforced focus on legal and judicial reforms is important in dealing with the challenges of prosecution of threats from cybercrime and cyberespionage. There is a need to implement and institutionalize strategies and policies to find effective ways for dealing with important challenges to the governance structure, organizations and institutions. This includes the need to offset problems related to fragmentation and repetition of functionalities, interoperability, challenges of multivarious (mis)scalability, standardization and problems from potential precise coordinated and targeted attacks as well as vulnerabilities that can overwhelm the defences of the State.

The fifth policy suggestion is related to *data/cyber sovereignty*. India should focus on demarcating the legality on aspects of sovereignty, jurisprudence, data retention, data preservation, data accessibility, data residency. Sovereignty, both cyber sovereignty and data sovereignty. An emphasis on creating "(potential) strategies and policies in the upcoming Science, Technology and Innovation (STI) policy... [and] in delegitimizing cyber sovereignty claims" by foreign digital transnational companies, private players and countries in Indian cyberspace/spatial paradigms with implications for Indian national security and citizens' privacy is a prerequisite. The stagnation and laggardness in promulgating India's data protection law or any legislation in the lines of qualities of Europe's General Data Protection Regulation (GDPR) are very important in defusing current and potential threats to sovereignty, legitimacy, control and (supervised) access of the Indian State over data from the country.

Further, careful implementation of recommendations from A.P. Shah Committee (2012), Justice B.N. Srikrishna Committee (2018), and that of the Supreme Court judgement in the K.S. Puttaswamy (Retd.) and Anr. Vs Union of India And Ors (2018) are required to deal with (potential) challenges. This will be a prerequisite in dealing with future Cambridge Analytica-type incidents, one of the biggest threats to Indian democracy. The government's focus should be on international, inter-governmental and/or multilateral cooperation, coordination and standardization. The utilization of mutual legal assistance treaties (MLATs) with other countries should be carefully enforced and maximized to circumvent issues or damages to foreign policy imperatives. Moreover, fragmentation of cyberspace, in the likes of multilateralism, digital globalization and digital sovereignty, is brewing further potential challenges to India's domestic and foreign policy. This requires the need to have new norms that adhere to balancing between "multilateral democratization" vs "securitization" in cyberspace, both in the domestic and international sphere.

CONCLUSION

Cyber attacks are a growing cause for concern for both State as well as non-State actors. It has been considered an extension of the traditional or conventional forces or actors, but taking place in the digital world or cyberspace. It is considered among the fastest-growing interaction and communication areas, with more and more actors entering the arena. It exploits the speed, convenience and anonymity for engaging in a wide range of activities that transcend the traditional "Westphalian" sovereign nature of physical borders. In terms of threats, it has acted as a medium through which significant harm has been caused, posing as a medium through which multiplier threats and victimization have occurred.

REFERENCES

Carayannis, E. G., & Campbell, D. F. J. (2009). 'Mode 3' and 'Quadruple Helix': Toward a 21st century fractal innovation ecosystem. *International Journal of Technology Management, 46*(3/4), 201–234.

Dilipraj, E., & Reghunadhan, R. (2018). Organisational governance of cyberspace in India. *Journal of Air Power and Space Studies, 13*(1), 115–134.

Federal Bureau of Investigation (FBI). (2018). *Cyber crime.* Retrieved June 15, 2019, from https://www.fbi.gov/investigate/cyber

Indo-Asian News Service (IANS). (2021). Nasscom partners with Microsoft, announces 'AI Gamechangers' programme. *ET Government.* Retrieved March 5, 2021, from https://government.economictimes.indiatimes.com/news/dig ital-india/nasscom-partners-with-microsoft-announces-ai-gamechangers-pro gramme/81306530

Langa, M. (2020). Minimum government, maximum governance: PM Modi's mantra to IAS probationers. *The Hindu.* Retrieved December 19, 2020, from https://www.thehindu.com/news/national/minimum-government-maximum-governance-pms-mantra-to-ias-probationers/article32990257.ece

PricewaterhouseCoopers (PwC). (2017). *Securing the nation's cyberspace.* Retrieved May 2, 2019, from https://www.pwc.in/assets/pdfs/publications/2017/securing-the-nations-cyberspace.pdf

Prime Minister's Office (PMO). (2021). *Digital India.* Retrieved *February 8, 2021,* from https://www.pmindia.gov.in/en/tag/digital-india/#skip_to_main

Reghunadhan, R. (2018). Cyber threat landscape of digital India: A critical perspective. *Journal of Polity and Society, 10*(1&2), 37–50.

Reghunadhan, R. (2021). Security-cum-privacy implications of whatsapp's privacy policy 2021: An indian perspective. *Science Policy Forum.* Retrieved

May 20, 2021, from https://thesciencepolicyforum.org/articles/perspecti
ves/security-cum-privacy-implications-of-whatsapps-privacy-policy-2021-an-
indian-perspective/

Sivanesan, S. (2018). Multi-disciplinary approach required for cyber-security.
Observer Research Foundation. Retrieved July 9, 2019, from https://www.orf
online.org/research/multi-disciplinary-approach-required-for-cyber-security/

Appendix I: Digital India Initiatives

- Aadhaar Enabled Payment System (AEPS)
- Aadhaar identity platform
- Agmarket portal or Agrimarket app
- Beti Bachao Beti Padhao
- Bharat Broadband Network
- Bharat Interface for Money (BHIM)
- Centre for Excellence for Internet of Things (COE-IT)
- Common Service Centre (CSC) scheme
- Consular Services Management System or "MEA" in Aid of Diaspora in Distress (MADAD) app
- Crime and Criminal Tracking Network & Systems (CCTNS)
- Crop Insurance Mobile App
- Cyber Swachhta Kendra (Botnet Cleaning and Malware Analysis Centre)
- Deen Dayal Upadhyaya Gram Jyoti Yojana
- Digidhan Abhiyaan
- DigiLocker
- DigiSevak or Volunteer Management System (VMS)
- Digital AIIMS
- Digital Saksharta Abhiyaan (DISHA) or National Digital Literacy Mission (NDLM) Scheme
- Digitize India Platform (DIP)
- Direct Benefit Transfer (DBT)

R. Reghunadhan, *Cyber Technological Paradigms and Threat Landscape in India*, https://doi.org/10.1007/978-981-16-9128-7

- Earth System Science Organization—Indian National Center for Ocean Information Services (ESSO-INCOIS)
- E-Basta
- E-Biz
- E-Courts
- E-District
- E-Granthalaya
- E-Greetings
- E-Hospital
- Election Commission of India Electronic Voting Machines (ECI EVM) Tracking
- Electronic Development Fund (EDF)
- Electronic Modified Special Incentive Package Scheme (E-MSIPS)
- Electronic National Agriculture Market (E-NAM)
- Electronic Transaction Aggregation & Analysis Layer (E-TAAL)
- Employees' Provident Funds Ordinance (EPFO) Web portal and mobile app
- E-Office
- E-Panchayat or Digital Panchayat
- E-Pathshala
- E-Prison
- E-Procurement Portal or Central Public Procurement Portal (CPP)
- E-Sampark
- E-Sign
- E-Tourist Visa
- E-Trade
- Farmer portal
- Fertiliser Monitoring System (FMS)
- Garv Grameen Vidyutikaran Mobile app
- Geographic Information System (GIS)
- Geological Survey of India (GSI)
- Goods and Service Tax Network (GSTN)
- Government e-Marketplace (GeM)
- Heritage City Development and Augmentation Yojana (HRIDAY)
- Himmat App
- ICDS Systems Strengthening and Nutrition Improvement Project (ISSNIP)
- India BPO Promotion Scheme (IBPS)

- Indian Railway Catering and Tourism Corporation (IRCTC) Connect
- Integrated Health Information Platform (IHIP)
- Jeevan Pramaan
- Khoya Paya portal
- Kisan Suvidha
- Knowledge Management System (KMS)
- KV Shaala Darpan
- Learning Management System (LMS)
- M-Asset
- MCessation Programme
- MeghRaj (formerly GI Cloud)
- Ministry of Corporate Affairs project (MCA_{21})
- M-Kavach
- M-Kisan SMS portal
- Mobile applications store (m-AppStore)
- Mother & Child Tracking System (MCTS)
- M-Rakt Kosh
- MyGov
- National Career Service Portal
- National Knowledge Network (NKN) project
- National Mission on Education using ICT
- National Scholarship Portal (NSP)
- National Super Computing Mission (NSM)
- National Ujala Dashboard
- National Voters Service Portal (NVSP)
- NIKSHAY online tool
- NIRBHAYA app
- North East BPO Promotion Scheme (NEBPS)
- NREGAsoft, OpenForge, PAHAL (DBTL)
- Online Labs (OLABS)
- Open Government Data (OGD) Platform India
- PARIVAHAN Portal
- Passport Seva Project (PSP)
- Paygov India or National Payment Service platform
- Pradhan Mantri Gramin Digital Saksharta Abhiyaan (PMGDISHA)
- Pradhan Mantri Jan-Dhan Yojana (PMJDY)
- Pradhan Mantri Kaushal Vikas Yojana (PMKVY)
- Program Management Information System (PMIS)

- Public Financial Management System (PFMS)
- Pusa Krishi app
- Rapid Assessment System (RAS)
- Saransh
- Shaala Sidhdhi or the National Programme on School Standards and Evaluation (NPSSE)
- Single Window Interface for Trade (SWIFT)
- Smart Cities Mission
- SMS-based Mid-Day Meal Monitoring Scheme
- Soil Health Card
- Startup India Portal and Mobile App
- Sugamaya Pustakalaya
- Sugamya Bharat Abhiyaan or Accessible India Campaign and Mobile App,
- Swachhta Abhiyan app
- SWAYAM
- Targeted Public Distribution System (TPDS)
- Udaan
- Unified Mobile Application for New-Age Governance (UMANG)
- Un-reserved Ticket through Mobile Applications (UTS App)
- Visvesvaraya PhD Scheme for Electronics and IT

Appendix II: Chronological Order of Policies, Legislations, Regulations and Reports in Dealing with the Issue of Cyber Technological Paradigms in India

Year	Policies/Legislations/Regulations/Reports
1885	Indian Telegraph Act
1933	The Indian Wireless Telegraphy Act
1981	National Frequency Allocation Plan (NFAP)
1994	National Telecom Policy
1995	Cable Regulation Act
1997	Satellite Communication Policy (SATCOM)
	Telecom Regulatory Authority of India Act
1999	New Telecom Policy
	Addendum To NTP
2000	Telecom Regulatory Authority of India (Amendment) Act
	Information Technology Act
2002	Universal Service Support Policy
2003	Indian Telegraph (Amendment) Act
	National Numbering Plan
2004	Broadband Policy
	New policy for.in internet domain name
	Indian Telegraph (Amendment) Rules
2006	Indian Telegraph (Amendment) Act
2008	Information Technology (Amendment) Act
2011	Information Technology (Guidelines for Cyber Cafe) Rules, 2011 (Privacy Rules)
	"Press Note" Technology (Clarification on the Privacy Rules)
2012	Report of "Group of Experts on Privacy" constituted by Planning Commission of India under chairmanship of A.P Shah

(continued)

© The Author(s), under exclusive license to Springer Nature
Singapore Pte Ltd. 2022
R. Reghunadhan, *Cyber Technological Paradigms and Threat Landscape in India*, https://doi.org/10.1007/978-981-16-9128-7

(continued)

Year	Policies/Legislations/Regulations/Reports
	Policy document "Recommendations of JWG on Engagement with Private Sector on Cyber Security"
	Preference for Domestically Manufactured Electronic Goods (PMA) Policy
	National Telecom Policy
2013	Guidelines for Securing the National Critical Information Infrastructures (NCII)
	National Cyber Security Policy (NCSP)
	National Policy on Universal Electronic Accessibility
2014	Policy on Cloud Computing- "Meghraj"- Reports on GI Cloud Strategic Direction Paper and GI Cloud Adoption and Implementation Roadmap
	Telecom Regulatory Authority of India (Amendment) Act
2015	Draft Policy on Internet of Things
	Indian Telegraph Act (Amendment) (August & December)
2016	National Encryption Policy
	Targeted Delivery of Financial and Other Subsidies, Benefits and Services Act (Aadhaar Act)
	Indian Telegraph Right of Way (RoW) Rules
	Radio Regulations
2017	Indian Telegraph Right of Way (Amendment) Rules
	Public Procurement (Preference to Make in India) Order
	Supreme Court Judgement in *Puttuswamy v. Union of India* on Fundamental Right to Privacy
2018	National Digital Communications Policy (NDCP)[1] (formerly National Telecom Policy)
	National Frequency Allocation Plan
	Report of the 5G High Level Forum by Steering Committee
2019	Report on non-personal data governance framework (the NPD Report)
	Personal Data Protection Bill 2019 (PDP Bill)
	National Broadband Mission
2020	Draft Space Communications (SPACECOM) Policy
	Norms, Guidelines and Procedures for Implementation of Space Based Communication Policy of India- 2020 (SPACECOM NORMS)
	Draft Space-based Remote Sensing Policy of India Policy (RS Policy)
	TRAI's Whitepaper on Smart Cities in India: Framework for ICT Infrastructure

(continued)

---

[1] There are three Mission envisaged by NDCP 2018 to achieve it objectives by 2022: *Connect India* to create "robust Digital Communications Infrastructure", *Propel India* to enable "Next Generation Technologies and Services through Investments, Innovation and IPR generation", and *Secure India* to ensure "Sovereignty, Safety and Security of Digital Communications" (DoT 2018).

(continued)

Year	Policies/Legislations/Regulations/Reports
2021	Draft Norms, Guidelines and Procedures for Implementation of RS Policy 2020 (RS NGP) Draft of National Strategy on Blockchain Information Technology (Intermediary Guidelines and Digital Media Ethics Code) Rules Guidelines for acquiring and producing Geospatial Data and Geospatial Data Services including Maps (Geospatial Policy)

Source Compiled by the Author

GLOSSARY

Access A process by which, users gain the rights to operate a local or remote system, application or program. The user may be required to enter an ID and password.

Algorithm An algorithm is a set of rules for getting a specific output from a specific input. It can be a program or a series of steps defining a modus operandi, which yields what is regarded to be an acceptable solution.

Anonymity It is related to privacy for the netizen for various activities in the internet.

Authentication The establishment of validity of a claimed identity of the user and its verification for any related access or activities thereupon.

Artificial Intelligence (AI) A term used to describe the use of a system to emulate human decision-making and learning abilities.

Bandwidth The rate at which data is transferred to, or from, a computer or appliance, using a medium that might be physical or wireless.

Bit A single indivisible item of binary data that might be '1' or '0'.

Bitcoin It is a kind of cryptocurrency based on decentralised distributed ledger system called Blockchain Technology.

Blackhat Hacker A hacker who attacks and breaks into a computer system, network or device(s) with malicious intent. It was coined by Richard Stallman, the founder of Free Software Movement.

Blockchain It is considered to have a consensual system of financial transactions and storage of information.

© The Author(s), under exclusive license to Springer Nature Singapore Pte Ltd. 2022
R. Reghunadhan, *Cyber Technological Paradigms and Threat Landscape in India*, https://doi.org/10.1007/978-981-16-9128-7

Botnet A collection of internet-connected devices that are infected and connected by a malware or remotely by another system or user.

Broadband A term used to describe access technologies and networks that typically offer bandwidths of 1 Mbps and more, thereby offering high-speed data transfer.

Browser A program used to scan and search electronic files and documents.

Chatterbot/ Chat Room A place for people to converse online by typing messages to each other. Once you are in a chat room, others can contact you by email. Some online services monitor their chat rooms and encourage children to report offensive chatter. Some allow parents to deny access to chat rooms altogether.

Client Server A distributed system of architecture where client systems are connected to server systems, providing an interface to applications and data that are stored on the server.

Compound annual growth rate It is the rate of return that would be required for an investment to grow from its beginning balance to its ending balance, assuming the profits were reinvested at the end of each year of the investment's lifespan.

Computer Network It means the interconnection of one or more computers through (i) the use of satellite, microwave, terrestrial line or other communication media; (ii) terminals or a complex consisting of two or more interconnected computers.

Computer System It means a device or collection of devices which contain computer programs, electronic instructions, input data, output data, that performs logic, arithmetic, data storage and retrieval, communication control, and other functions.

Computer A system or appliance that is able to process information based on the input provided by the user and generates results.

Content It includes text, images, sounds, videos, and animations. Also called Web content, internet content.

Cryptocurrency It is digital currency, with encryption tools and techniques used for its generation; eg: Bitcoin, Ethereum, Ripple, Litecoin, NEO, Stellar

Cryptography A process that ensures that data or information is read or used only by its intended readers or users. It includes encryption (disguising input information or data) and decryption (returning the data to original, usable form).

Cyber Security Intelligence Index An annual report brought out by IBM detailing security challenges and emerging vulnerabilities in various computer systems and devices.

Cyberspace It is another name for the virtual medium, and was coined by William Gibson in his book *Neuromancer* (1986).

Data A representation of information in a manner that is suitable for communication, interpretation, storage and processing.

Database An electronic information system offering data, storage and its retrieval.

Decryption A process by which encrypted data is unlocked to become readable.

Defense Critical Infrastructure It refers to the composite of military and non-military assets essential to project, support, and sustain military forces and operations worldwide.

Defense Industrial Base It refers to the Department, Government, and private sector worldwide industrial complex with capabilities to perform research and development, design, produce, and maintain military weapon systems, subsystems, components, or parts to satisfy military requirements.

Deep Learning A sub-field of machine learning, and is influenced by artificial neural networks.

Digital A term describing devices, such as a computer, that process and store data in the form of ones and zeros.

Digitalisation It is the transformation that occurs in various sectors, particularly being connected to the internet and cyberspace.

Domain Name Service (DNS) A server that converts domain names into IP addresses.

Downloading The process of copying filed from a remote server to a local computer.

Electronic commerce (e-commerce) It includes both indirect e-commerce, the electronic ordering of tangible goods that are physically delivered through traditional channels and direct electronic commerce, i.e., online-ordering, payment and delivery of intangible goods and services such as computer software, entertainment content or information services on a global scale.

Email A method of communicating documents and digital files electronically; a computer-based equivalent of a letter.

Encryption The process of ciphering messages or data so that it may be deciphered and read only by the intended recipients.

Ethical Hacking It is related to hacking activities intended to understand and point-out vulnerabilities and security gaps in the defense measures, mechanisms and procedures undertaken in a computer system and network.

Hacking A term used to describe an individual (hacker) who endeavours to gain unauthorised and often illegal access to a computer system(s), devices and/ or networks.

Homepage A file on the internet that contains information and links to other files (or processes) and locations on the Internet.

Hypertext Transfer Protocol (http) A standard protocol that allows web browsers to communicate with web servers.

Information Data that have become meaningful as a result of collection, processing, organisation and interpretation.

Integrity The property that data or information has not been modified or altered in an unauthorized manner.

Internet Service Provider (ISP) A service that allows the user to connect to the World Wide Web (www). When the user signs up (it takes special software and a modem), he will be asked to enter a login name and a secret password.

Internet The universal network that allows computers to talk to other computers in words, text, graphics and sound anywhere in the world.

Interoperability Interoperability of cryptographic methods means the technical ability of multiple cryptographic methods to function together.

Key A sequence of easily changed symbols that, used with a cryptographic algorithm, provides a cryptographic process.

Knowledge An organised body of information representing a description and understanding of concepts and relationships.

Localisation It is a highly technical process by which computer programs written in one language by members of one culture are translated into another language for use by members of another culture.

Links Highlighted words on a website that allow the user to connect to other parts of the same website or to other websites.

Machine Learning It is the process in which machines are provided techniques and tools to act on their own and take decisions independent of human intervention.

Malware It is a software designed to disrupt, damage and/ or steal data as well as information.

Metadata A term used to describe data that indicates the information types and subjects. In the context of the Web, metadata such as indexes and URLs are gathered and stored by search engine implementations.

Modem An internal or external device that connects your computer to a phone line.

Navigate The process of moving between points via links in hypertext-based material such as web applications.

Netizen An internet user, who is accustomed in getting information.

Neural Networks (Artificial) A computer system modelled on the functioning of the human nervous system.

Online Service An ISP with an added information, entertainment and related features.

Operating System (OS) A generic term used to describe the software elements that manage system resources and thus provide an interface between the user and the system, as well as between software and the system. Popular OS are Windows XP, Windows 10, Unix, IOS, MS-DOS.

Password A series of alphanumeric characters used to protect a system against unauthorised access.

Peer-to-Peer Network A network that permits each network user to access the directories and peripheral devices associated with any connected computer. eg: Peer-to-Peer Electronic Cash System, Peer-to-Peer Bitcoin Network.

Protocol A format used to transmit and to receive data like IP, SMTP, HTTP.

Ransomware It is a malicious software that intrudes, attacks and denies access to a device or system until ransom is paid for.

Real-Time A program or system that responds to user interaction instantly or captures data at the rate it actually occurs.

Search Engine A facility that lets the user search for information by typing a few keywords. It retrieves the document in which specified keywords or phrases are found.

Sovereignty Supreme and independent power or authority in government as possessed or claimed by a state or community.

Spam A form of unsolicited email. It is the internet equivalent of junk mail.

Spammed The act being swindled using spam mails.

Territorialization The act of organizing as a territory.

Territory An area of land or sea, that is considered as belonging to or connected with a particular country or State actor.

Trojan It is a 'virus-laden' program which encourages the user to run it but its real purpose is to damage files on the system.

Uniform Resource Locator (URL) The address of a service or website or web page, which can be used by the web browser to open specific sites and pages.

Very Small Aperture Terminal (VSAT) Network It is a relatively small earth-sized satellite communications station that transmits data, video and voice signals.

Virus 'Vital Information Resources Under Seige' is a file that is maliciously planted in a computer that can damage files and disrupt the system.

Web Server An architecture which maintains the connection between the processing and data of the server-side with that of the client-side.

Website (Web Site) It is a destination a user can look at and retrieve data from. All the websites in the world, linked together, make up the World Wide Web or the 'Web'.

Whitehat Hacker A computer specialist who breaks into systems and networks to assess security vulnerabilities.

Windows An industry-standard graphical user interface (GUI) and OS for the PC platform developed by Microsoft.

Worm (Computer) A computer worm is a type of malicious software program whose primary function is to infect other computers while remaining active on infected systems.